Stochastic Methods for Parameter Estimation and Design of Experiments in Systems Biology

Von der Fakultät Konstruktions-, Produktions- und Fahrzeugtechnik der Universität Stuttgart zur Erlangung der Würde eines Doktor- Ingenieurs (Dr.-Ing.) genehmigte Abhandlung

Vorgelegt von

Andrei Kramer

aus Novokusnezk

Hauptberichter: Prof. Dr. Nicole Radde
Mitberichter: Prof. Dr. Johannes Kästner
Prof. Dr. Tim Beissbarth

Institut für Systemtheorie und Regelungstechnik der Universität Stuttgart

2015

Bibliographic information published by the Deutsche Nationalbibliothek

The Deutsche Nationalbibliothek lists this publication in the Deutsche Nationalbibliografie; detailed bibliographic data are available on the Internet at http://dnb.d-nb.de .

ISBN 978-3-8325-4195-8

Logos Verlag Berlin GmbH
Comeniushof, Gubener Str. 47,
10243 Berlin
Tel.: +49 (0)30 42 85 10 90
Fax: +49 (0)30 42 85 10 92
INTERNET: http://www.logos-verlag.de

Acknowledgement

For their great camaraderie and support I would like to thank all members of the *Institute for Systems Theory and Automatic Control* (IST). Some graduated and left, some new members joined the group, the composition was in slow but steady flow. Yet, what didn't change was the general spirit of the group, the enthusiasm for research and moral support of each other.

By far the most support I have received from my adviser Prof. Dr. Nicole Radde, for which I am immensely grateful. She was supportive without fail, invited discussions at all opportune moments and provided helpful, critical feedback very frequently. She also suggested to visit Prof. Mark Girolami's group, as he has done a lot of research on the topic of this dissertation. This research stay at UCL's statistics department turned out to be incredibly beneficial and influential on my work. It resulted in two collaboration works with members of the department. I am very grateful to both Nicole Radde and Mark Girolami for this opportunity and all statistics group members for the warm welcome, our discussions, chats and friendship. Specifically, I would like to thank Dr. Ben Calderhead and Dr. Vassilios Stathopoulos as they have invested the most time and effort in our respective collaborations. I also thank the SimTech graduate school (GS SimTech) and German Research Foundation (DFG) for financial support of this project within the Cluster of Excellence in Simulation Technology (EXC 310/2) at the University of Stuttgart, which also made this research stay possible. I thank both additional members of my examination comittee, Prof. Dr. Tim Beißbarth and Prof. Dr. Johannes Kästner for reading the dissertation, for investing some of their time to join the defence presentation and for our discussion afterwards.

Many of the IST members were not only dear colleagues but over time became cherished friends. I would like to thank especially Christian Breindl, Jan Hasenauer and Patrick Weber for an enormous amount of help, great fun, intense discussions and proof reading of many parts of this manuscript. In the same spirit, I would like to thank Alexander Rabe and my brother Wladimir Kramer for further proofreading. I am also very grateful to my wife Alina Miehe for her love and support.

Contents

List of Figures

List of Tables

Deutsche Kurzfassung

Die Systembiologie untersucht biochemische Reaktionsnetzwerke mit dem Ziel, die in den Zellen stattfindenden Prozesse durch eine Kombination von Experimenten und mathematischer Modellierung zu verstehen und vorhersagbar zu machen.

Innerhalb von Zellen (Eukaryoten und Prokaryoten) ereignen sich Reaktionsketten zwischen komplexen organischen Molekülen. Diese Reaktionsketten bilden teilweise Netzwerke, denen man bestimmte Funktionen, wie Signalübertragung oder Messung von Zyklen zuordnen kann. Die Reaktanten lassen sich in drei Gruppen unterteilen: Substrate (Ausgangsstoffe), Produkte und Modifikatoren. Modifikatoren katalysieren Reaktionen, die sonst nur sehr selten stattfinden würden; alternativ können sie Reaktionen hemmen oder komplett unterbinden. Sie werden bei der Reaktion, im Gegensatz zu Substraten, nicht verbraucht. Stoffklassen, die wir betrachten, umfassen unter anderem Proteine, Lipide, Zucker, DNA und RNA. Eine robuste und Ausführung intrazellulärer Prozesse wird mithilfe von Rückkopplungen (Feedback-loops) realisiert, welche den Zweck einer *Regelung* verfolgen. Solche Rückkopplungsmechanismen machen es schwer, das Gesamtverhalten eines Reaktionsnetzwerks rein durch Intuition nachzuvollziehen. Aus diesem Grund werden in der Systembiologie mathematische Modelle zur Auswertung von Experimenten verwendet, mit dem Ziel quantitative oder qualitative Vorhersagen zu treffen.

Häufig modellierte und zum Teil schon recht gut verstandene Prozesse bilden Signalübertragungsketten, welche mit dem Binden eines Liganden an spezifische Rezeptoren in der Zellmembran aktiviert werden. Das Signal wird durch Aktivierungskaskaden in das Zellinnere weitergegeben. Dies stellt einen möglichen Ansatzpunkt für die Modellierung eines *Eingangssignals* dar. Das Resultat der Reaktionskaskade könnte beispielsweise ein geändertes Expressionsmuster von Proteinen in der Zelle sein. Die messbaren Eigenschaften einer so veränderten Zell-Funktion werden dann mit einem Modell verglichen, welches einem vergleichbaren Eingangssignal ausgesetzt wird. Für die Konstruktion des mathematischen Modells ist es hilfreich, alle beteiligten Reaktanten zu kennen. Die Menge und Qualität

		Zustandsvariablen	
		kontinuierlich	diskret
Zeit	kont.	gewöhnliche DGL; partielle DGL; stochastische DGL	chemical master equation; Gillespie Algorithmus
	diskret	autoregresive Modelle: AR, VAR, ARMAX; Bayes'sche Netze	Boolesche Netze; Daten mit geringer Auflösung

Tabelle 1.: Stichwortsammlung zu den 4 Systemklassen in der Systembiologie. Gewöhnliche Differentialgleichungen (DGL) stehen im Fokus dieser Arbeit.

von Vorwissen bestimmen dabei häufig die Wahl der Modellklasse. Wir unterteilen Modelle grob in vier Klassen: (1) zeit- und zustandsdiskret, (2) zeitdiskret, zustandskontinuierlich, (3) zeitkontinuierlich, zustandsdiskret (4) zeit- und zustandskontinuierlich. Tabelle 1 listet stichwortartig typische Modelltypen für diese Klassen auf. Lösungsverfahren für die mathematischen Aufgabenstellungen dieser Ansätze verwenden mitunter eine endliche Anzahl rationaler Stützstellen für Variablen, die im Modell kontinuierlich sind. Die Lösung wird also gewissermaßen durch Diskretisierung angenähert. Der Feinheitsgrad dieser Diskretisierung ist jedoch fast beliebig einstellbar[1], was bei fundamental diskreten Modellen wie Booleschen Netzen nicht ohne weiteres möglich ist.

Auf oberster Abstraktions-Ebene nehmen wir eine

`Eingang`→`System`→`Messung`

Sicht an. Unabhängig von der Modellklasse sind die zugrunde liegenden Modelle von Eingang, System und Ausgang in den allermeisten Fällen parametrisiert und es gilt Parameter zu finden, die alle bisher gesammelten Beobachtungen mithilfe des Modells reproduzieren können.

Thema und Beitrag dieser Arbeit

Wir beschränken uns auf den Fall von vollständig kontinuierlichen Systemen (Klasse 4) und wählen als Werkzeug der Parameteridentifizierung die

[1] `rk23` → `rk45`

Bayes'sche Modellanalyse. Diese Methode weist allen Parametern $\theta_i \in \mathbb{R}$ $(i = 1, \ldots, m)$, welche als unsicher eingestuft werden, eine Wahrscheinlichkeitsdichte zu, die den Wissensstand über den Wert des Parameters darstellt. Die Parameter sind in ihrer Wirkung oft stark miteinander korreliert. Dies ist beispielsweise der Fall für Parameterpaare mit gegensätzlichen Rollen wie Abbauraten und Syntheseraten, aber auch bedingt durch die Netzwerkstruktur der Interaktionen. Aus diesem Grund ist der Formalismus grundsätzlich darauf ausgelegt, unter Berücksichtigung der Daten $\mathcal{D}$ und des Modells M eine multivariate *a-posteriori* Verteilung $p(\theta|\mathcal{D}, M)$ zu finden, welche eine solche Gesamtwirkung des Parametervektors $\theta \in \mathbb{R}^m$ beschreibt. Nicht immer ist die Datenmenge ausreichend alle Parameter zu identifizieren; in manchen Fällen verhindert die Struktur von System und Messausgang eine eindeutige Schätzung der Parameter (strukturelle Nichtidentifizierbarkeit). Hinzukommt, dass die Parameter auch nach einem Modell-Fit allein durch das Messrauschen der Daten unsicher bleiben. Die Hoffnung dabei ist, dass die Unsicherheit gegenüber der a-priori Verteilung $p(\theta|M)$ sinkt, also nach Berücksichtigung der Daten. Diese Unsicherheit muss sich zwingend auf Vorhersagen auswirken, sie kann reduziert, aber niemals beseitigt werden.

In unserem Fall ist eine Vorhersage charakterisiert durch ein Anfangswertproblem. Insbesondere ist das Thema dieser Arbeit geprägt durch Modelle, deren innere Struktur komplex genug ist um eine geschlossene Lösung sehr schwierig zu machen. Infolgedessen ist auch die a-posteriori Verteilung nicht analytisch zugänglich, sie ist nicht direkt (analytisch) integrierbar. Die charakteristischen Eigenschaften der Parameter-Dichte wie Mittelwert, Kovarianz und andere Momente lassen sich nicht exakt berechnen und darüber hinaus können wir auch keine analytische Transformation der Parameterunsicherheit in den Zustandsraum des Modells ausführen.

Zusammengefasst bedeutet dies, dass Vorhersagen nicht durch analytische Erwartungswerte gemacht werden können, sondern nur durch numerische Verfahren. Ein sehr weit verbreitetes Vorgehen ist es, mithilfe von *Markov chain Monte Carlo* (MCMC) eine repräsentative Stichprobe von Parametervektoren Θ zu erzeugen. Jeder einzelne dieser Vektoren kann anschließend dazu benutzt werden, durch Vorwärtssimulation des Modells eine Vorhersage zu treffen. Kumulativ, mit einer passenden Stichprobengröße, bilden diese Vorwärtssimulationen eine statistische Vorhersage des Systemverhaltens zu jedem vorgegebenen Eingangssignal. Wir erhalten damit eine Aussage, die untrennbar mit einer Unsicherheitsanalyse verbunden ist. Für den Einsatz von MCMC ist es lediglich notwendig, Werte von $p(\theta|\mathcal{D}, M)$ berechnen zu können. Das erfordert in unserem Fall den Satz von Bayes.

Die Anwendung von MCMC Algorithmen leidet jedoch unter dem eher schlechten Konvergenzverhalten von MCMC Schätzern. Die Konvergenzrate von Funktionen der Samplingvariablen θ mit der Stichprobengröße N ist $N^{-\frac{1}{2}}$. Die benötigte Stichprobengröße erhöht sich zudem exponentiell mit der Dimension des Problems. In unseren Beispielen erstrecken sich die Sampling-Zeiten von wenigen Minuten bis hin zu einigen Tagen. MCMC Algorithmen, welche in einer gegebenen Zeit (effektiv) größere Stichproben liefern, sind deswegen wünschenswert.

In dieser Arbeit beschreiben wir eine Reihe verschiedener MCMC Algorithmen und vergleichen ihre Effizienz. Wir stellen ein Softwarepaket vor, welches eine dieser Methoden in der Programmiersprache `C` implementiert. Darüber hinaus beschreiben wir von uns entwickelte Varianten einiger MCMC Algorithmen vor, welche unter Ausnutzung bestimmter Eigenschaften der Daten bedeutend schneller sind als ihre Grundversionen. Effizienz bedeutet in diesem Zusammenhang, in kürzerer Zeit a-posteriori Parameter-Stichproben zu liefern, die dabei wenig Auto-Korrelation zwischen den Markov-Ketten-Gliedern aufweisen. Hierzu definieren wir die effektive Geschwindigkeit einer Markov Kette.

Für Daten, die als Zeitreihe vorliegen, haben wir die Algorithmen HMC (Hamiltonian Monte Carlo[2]) und SMMALA (simplified manifold Metropolis adjusted Langevin algorithm) modifiziert. Wir konstruieren eine Abschätzung für die Parameter Sensitivität $S_f(t) = \partial_\theta x(t)$ der Trajektorien, welche besonders akkurat ist, wenn die Modell-Trajektorien monoton zu einem stabilen Fixpunkt konvergieren. Diese Abschätzung ist bedeutend leichter zu berechnen als eine vollständige Sensitivitätsanalyse des Modells und erleichtert die Berechnung der Metrik parametrisierter Wahrscheinlichkeitsdichten ungemein. Wir nennen diese Variante SSMMALA; der Algorithmus ist, unter Berücksichtigung der Auto-Korrelation, 18(5) mal schneller als SMMALA bei einer Modellgröße von $n = 6$ Zustandsvariablen und $m = 9$ Parametern.

Falls das System sich während der Messung in einem stabilen Fixpunkt befindet, berechnen wir die Fixpunkte des Modells direkt, ohne die transiente Phase zu simulieren. Wir benutzen dazu, falls möglich, den *circuit breaking algorithm* CBA, eine Graph-theoretische Methode, die Fixpunkte besonders effizient berechnen kann. Der CBA ermöglicht es in einigen Fällen, ein Fixpunkt-Problem in eine Nullstellensuche von Polynomen umzuwandeln. In unserem Beispiel war diese Umformung möglich und die neue Variante CBA+SMMALA war gegenüber der Grundform (SMMALA) etwa 200 mal schneller. Modelle,

[2] auch Hybrid Monte Carlo

die diese Umformung nicht zulassen, profitieren dennoch von der Fixpunkteigenschaft der Daten. Wir haben die exakte, analytische Sensitivitätsanalyse von Fixpunkten mit dem Newton-Raphson Verfahren für Nullstellensuche kombiniert und die beiden Varianten NR+RMHMC und NR+SMMALA genannt. Auch hier beobachten wir eine deutliche Beschleunigung: bei einem Modell mit $n = 6$ Zustandsvariablen und $m = 14$ Parametern ist NR+RMHMC mehr als 1930(290) mal schneller als das zugrunde liegende RMHMC, während NR+SMMALA bei dieser Problemgröße etwa 10(4) mal schneller ist als die Grundform SMMALA.

Typischerweise werden neue Algorithmen in Skript-Sprachen wie MATLAB oder PYTHON implementiert. Diese Sprachen erleichtern die *Entwicklung* und die *Anpassung der Algorithmen durch den Benutzer* gegenüber einer Programmiersprache wie `C` oder `fortran` enorm. Skriptsprachen sind aber gleichzeitig deutlich langsamer in der Programm-Ausführung. Wir haben uns entschieden, den etablierten SMMALA Algorithmus komplett in `C` zu implementieren und beobachten eine schnellere Ausführung um den Faktor 4000 gegenüber einer äquivalenten Skript-Variante (in GNU OCTAVE).

Im folgenden Abschnitt geben wir einen Überblick über die Gliederung dieser Arbeit und fassen den Inhalt zusammen.

Gliederung und Zusammenfassung

Wir beginnen mit einer Auflistung der von uns verwendeten Konventionen und Symboltabellen. Da der Fokus dieser Arbeit eng mit der Angabe von Unsicherheiten verbunden ist, wollen wir hier auf die zwei wichtigsten Vereinbarungen hinweisen: (1) Wenn eine Zahl weder von uns vorgegeben ist, noch ein explizit angegebenes Fehlermodell hat, gehen wir von einem normal-verteilten Fehler aus und geben die Unsicherheit (1σ) in Klammern an (*concise notation*); (2) unser Ziel ist immer eine *Größtfehlerabschätzung*. Bei einer Fehlerfortpflanzung von einem Argument auf einen Funktionswert genügt uns eine Abschätzung nach oben für den Fehler des Funktionswertes. Dieses Kapitel schließt mit einer Beschreibung der Notation im Hinblick auf Statistik und Wahrscheinlichkeit.

Anschließend folgt ein Kapitel zum Hintergrund der mathematischen Modellierung. Darin beschreiben wir einige der vorhandenen Modellklassen in der Systembiologie und verwandten Disziplinen wie der numerischen Molekulardynamik (MD), Physik und Biochemie. Bei der Entscheidung für eine Modellklasse spielt die gewünschte Modellierungsskala, die Wahl der Zustandsvariablen und das Auflösungsvermögen der Messapparatur eine große Rolle. Die Hauptaussage dieses Kapitels ist, dass Modelle eine gute Balance zwischen Approximation und fundamentalen Grundlagen halten müssen, um hohe Vorhersagekraft zu haben. Dies beendet unsere Vorbetrachtungen.

Kapitel 2 – Einführung

Hier definieren wir die Komponenten der von uns verwendeten Modelle: Anfangswertprobleme von gewöhnlichen Differentialgleichungen. Wir definieren an dieser Stelle auch nützliche, abgeleitete Größen wie Zustands-Sensitivitäten.

Bei der Definition gehen wir insbesondere auf typische Situationen bei der Verwendung von Western Blot Daten ein und stellen vor allem den Datentyp der Fixpunkte (Gleichgewichtslagen) vor. Die von uns entwickelten Variationen bekannter MCMC Algorithmen (HMC & SMMALA) nutzen die besonderen Eigenschaften dieser Art von Daten aus, um die Erzeugung von Parameterstichproben drastisch zu beschleunigen.

Wir gehen auch auf Methoden der Parameterschätzung und deren Zusammenhang mit der Unsicherheitsanalyse des resultierenden Modell-Fits ein. Wir führen das von uns verwendete Modell der Messfehler ein und leiten daraus mithilfe des Satzes von Bayes die Struktur der a-posteriori Verteilung her. Sie setzt sich insbesondere aus einem *Likelihood* Term und der *a-priori* Verteilung zusammen. Die Likelihood gibt an, wie plausibel die Daten unter einer vorgegebenen Modell-Parametrisierung sind. Die a-priori Verteilung stellt einerseits das Vorwissen über mögliche Parameterwerte dar und kann andererseits auch zur Regularisierung des Problems verwendet werden. Optimierungsverfahren erwarten oft ein beschränktes Gebiet (eine konvexe Hülle) in dem die Optimierung stattfinden soll. Eine flache (uniforme) a-priori Verteilung ist das direkte Analogon zu dieser Beschränkung, während eine Normalverteilung dem Suchbereich weiche Grenzen setzt.

Kapitel 3 – Bekannte MCMC Methoden

Markov chain Monte Carlo Methoden stellen eine Kombination von Monte Carlo Verfahren und der Theorie der Markov Ketten dar. Wir geben eine kurze Einführung in beide Themen und erklären daraufhin die Funktionsweise der bekanntesten MCMC Verfahren. Diese lassen sich grob in drei Klassen unterteilen: (1) Metropolis-Klasse, (2) Hamilton-Klasse, (3) Langevin-Klasse.

Alle drei lassen sich mit dem Verfahren von *delayed rejection* kombinieren. Die Markovketten dieser Verfahren generieren eine Folge von Parameter-Vektoren: $\theta_0 \mapsto \theta_1 \mapsto \theta_2 \mapsto \ldots \mapsto \theta_N$. Dieser Prozess erfolgt iterativ und lässt sich in zwei Schritte aufteilen. Zuerst erfolgt ein *Vorschlag* für einen Nachfolger, der anschließend akzeptiert oder verworfen wird. Die drei von uns verwendeten MCMC Klassen sind wie folgt charakterisiert:

1. Zu einem gegebenen Wert $\theta_i \in \mathbb{R}^m$ wird ein Nachfolger $\phi \in \mathbb{R}^m$ zufällig vorgeschlagen. Dieser wird akzeptiert ($\theta_{i+1} = \phi$) oder verworfen ($\theta_{i+1} = \theta_i$).

2. Ein Nachfolger ϕ entsteht durch die Konstruktion einer Trajektorie im Raum der Parameter durch das Hamilton'sche Bewegungsgesetz mit zufälligen Anfangsbedingungen. Der negative Logarithmus der a-posteriori Verteilung wird dabei als Potential benutzt. Auch hier kann der Nachfolger unter Umständen verworfen werden, jedoch deutlich seltener als bei den anderen Verfahren.

3. Verfahren der dritten Klasse sind in gewisser Hinsicht eine Mischung der beiden Vorgänger. Der Vorschlag eines Nachfolgers ϕ zu einem gegebenen Punkt θ_i wird zufällig aus einer Normalverteilung gezogen. Der Mittelwert μ dieser Verteilung wird deterministisch aus θ_i und dem Gradienten der Ziel-Verteilung erzeugt. Diese Vorgehensweise entspricht der Zeit-Diskretisierung einer stochastischen Differentialgleichung, der Langevin Gleichung.

Kapitel 4 – Algorithmus Tuning

Alle zuvor vorgestellten Algorithmen haben freie interne Parameter, wie Schrittweiten oder die Länge der Trajektorien bei der numerischen Integration der Hamilton'schen Bewegungsgesetze. Dieses Kapitel diskutiert das Problem, diese *Parameter* für einen Vergleich zweier Algorithmen gleichwertig *einzustellen* (tunen). Gleichwertig bedeutet in dem Fall, dass keiner der beiden Algorithmen allein durch eine schlechte Parametrisierung schlechter abschneidet als der Konkurrent.

Wir stellen ein mögliches Verfahren vor, um zwei sehr unterschiedliche MCMC Methoden passend zu parametrisieren und damit vergleichbar zu machen.

Kapitel 5 – Neue Varianten der diskutierten Algorithmen

Unter Verwendung von analytisch berechneten Fixpunkt-Sensitivitäten lassen sich Gradienten des Logarithmus der a-posteriori Verteilung schneller berechnen. Diese Gradienten spielen eine wichtige Rolle in den Hamilton und Langevin Klassen der MCMC Algorithmen. In diesem Kapitel stellen wir von uns entwickelte MCMC Varianten vor, die speziell bei der Verwendung von Fixpunkt-Daten (Messungen von Gleichgewichtslagen) deutlich schneller arbeiten als ihre allgemeineren Grundversionen. Wir geben die Ergebnisse unserer Vergleichsstudien tabellarisch an und stellen die dabei erzeugten Stichproben graphisch dar. Wir verdeutlichen die Nützlichkeit der Bayes'schen Modellanalyse indem wir Vorhersagen über mögliche Experimente treffen. Diese Modell-gestützte Auswahl eines Experiments zielt darauf ab die verbleibende Parameterunsicherheit so stark wie möglich zu reduzieren.

Kapitel 6 – Softwareentwicklung in C

Wir haben ein bewährtes MCMC Verfahren (SMMALA) ausgewählt und ein solches Software-Paket geschrieben: `mcmc_clib`. In diesem Kapitel beschreiben wir die Funktionsweise und den Entwicklungsstand der Software. Der wohl größte Nachteil dieses Ansatzes ist, dass das resultierende Programm in gewisser Weise unflexibel ist und der Benutzer nur sehr schwer Strukturen wie *Likelihood* oder das zugrunde liegende Messmodell ändern kann. Der Geschwindigkeits-Vergleich mit OCTAVE Skripten ist jedoch sehr vielversprechend.

Kapitel 7 – Schlussfolgerungen und Ausblick

Die Anpassung der MCMC Algorithmen an Fixpunktdaten war sehr erfolgreich und ergab eine Beschleunigung gegenüber der jeweiligen Grundversion von bis zu zwei Größenordnungen. Auch der Vorteil der C Implementierung gegenüber Skript-Sprachen (hier GNU OCTAVE und MATLAB) war enorm: 3 Größenordnungen.

Wir beenden dieses Kapitel mit einem Ausblick auf eine mögliche Weiterentwicklung der C Software.

Nomenclature and Conventions

Some variables will retain their meaning over the entire length of this manuscript. Here we present a table of these variables, structured by topic. Others will take on different meanings depending on the context (statement, equation or section); these are sometimes called *bound variables* or *dummy variables*.

For brevity of notation, we will assign the same variable name to different quantities which nevertheless represent the same concept or entity, e.g. $x_i(t_j;\rho,u_k) =: x_{ijk}$; on the left x is a function (the solution to an initial value problem), on the right x is a three dimensional array. This is sometimes called *overloading* a name. In addition, we will often omit the full list of arguments for the solution function $x(t;\rho,u)$, especially when it is used as the argument of another function with very similar arguments: $h(x_i(t_j;\rho,u_k),\rho,u_k) =: h(x_{ijk},\rho,u_k)$, so x_{ijk} was indeed calculated using ρ. The use of a semicolon «;» in function arguments is equivalent to a comma and is meant to emphasise the primary argument to a function as opposed to (less often changing) function *parameters*.

Sets of Numbers

Symbol	Description
$\mathbb{N}$	natural numbers
$\mathbb{Z}$	integers
$\mathbb{R}$	real numbers
$\mathbb{R}_+$	nonnegative real numbers
$\mathbb{R}_{++}$	strictly positive real numbers, $\mathbb{R}_+ \setminus \{0\}$

Ordinary Differential Equations

Symbol	Domain	Description
n_a	$\mathbb{N}$	the size of a vector or list, with the name of the vector indicated by the slanted index a
t	$\mathbb{R}$	time
T	$\mathbb{N}$	number of measurement times
t_j	$\mathbb{R}$	a measurement time instance, $j = 1, \dots, T$
$\mathcal{E}_k$		an experiment; contains experimental conditions and measurement model, $k = 1, \dots, n_E$
u_k	$\mathbb{R}^{n_u}$	experimental conditions, or equivalently *inputs* for initial value problems in experiment $\mathcal{E}_k$
θ	$\mathbb{R}^m$	logarithmic parameters of ordinary differential equation models, primary Markov chain variables
ρ	$\mathbb{R}^m_{++}$	(strictly positive) parameters of ordinary differential equation models
φ		flow associated with the right hand side of an ordinary differential equation, $\varphi: \ t, \rho, u, x_0 \mapsto x(t, \rho, u)$
$x(t, \rho, u_k)$	$\mathbb{R}^n$	a specific solution to an initial value problem (under implicit initial conditions)
x_{ijk}	$\mathbb{R}$	values of $x_i(t_j; \rho, u_k)$; in this context x has the size $n \times T \times n_E$
x_{jk}	$\mathbb{R}^n$	values of $x(t_j; \rho, u_k)$
$\bar{x}_k$	$\mathbb{R}^n$	k-th steady state of $x(t)$
$\partial_x f(x)$		$\frac{\partial f(x)}{\partial x}$, with x being a formal argument
$J_f(x; \dots)$	$\mathbb{R}^{n \times n}$	the Jacobian of f. This is identical to: $\partial_s f(s; \dots)\rvert_x$

Statistics

Symbol	Domain	Description
$p(\rho)$	$\mathbb{R}_+$	probability density of ρ
$p(\rho \vert y)$	$\mathbb{R}_+$	probability density of ρ given y
$P(\omega)$	$[0, 1]$	probability of event ω
$P(\omega_1 \vert \omega_2)$	$[0, 1]$	conditional probability of event ω_1 given that ω_2 occurred
$\langle \rho \rangle$		expected value of ρ with respect to the posterior probability distribution

$[\rho]$	Monte Carlo estimate of $\langle\rho\rangle$ using a sample
$[\rho; N]$	Monte Carlo estimate of $\langle\rho\rangle$ using a sample of size N, given explicitly

Algorithm Names

HMC	Hamiltonian Monte Carlo
RMHMC	Riemannian manifold HMC
SMMALA	simplified manifold Metropolis adjusted Langevin algorithm
NR+SMMALA	Newton Raphson method combined with SMMALA
NR+HMC	Newton Raphson method combined with HMC
NR+RMHMC	Newton Raphson method combined with RMHMC
DRM	delayed rejection Metropolis
DRAM	delayed rejection adaptive Metropolis

General Operators

Symbol	**Domain**	**Description**
$v_{\cdot jkl}$	$\mathbb{R}^{n_v}$	index operation; the result is a vector w with entries $w_i = v_{ijkl}$, i.e. $w = v_{\cdot jkl}$ (this is similar to MATLAB's : operator)
$v_{i<j}$	$\mathbb{R}^{j-1}$	this index operation returns a *shortened* vector w which is entry wise equal to v up to $i = j$: $w_i = v_i$ $(i = 1, \dots, j-1)$
$\lceil r \rceil$	$\mathbb{R} \to \mathbb{Z}$	`ceiling` function, returns the smallest integer not smaller than r
$\mathrm{diag}(v)$	$\mathbb{R}^{n_v} \to \mathbb{R}^{n_v \times n_v}$	if v is a vector $V = \mathrm{diag}(v)$ is a diagonal matrix with $V_{ii} = v_i$
δ_{ij}	$\{0, 1\}$	Kronecker delta (returns 1 if $i = j$, 0 otherwise)
$A \backslash b$		solve linear equation $Ax = b$
$\|x\|$	$\mathbb{R}_+$	norm of x

Notation of Uncertainty

There are few exceptions to the following rule: a number, that is not an input to the problem, given without a *measure of uncertainty* is useless. Uncertainty can be quantified by a fixed upper and lower bound, a standard deviation, variance or estimate of a measuring device's systematic error. In this spirit, we choose to closely connect numbers to their uncertainty by using the following notation for errors throughout this manuscript:

$$v = 1.71(12) := 1.71 \pm 0.12 \qquad a = 42(3) \times 10^6 := (42 \pm 3) \times 10^6 \,, \tag{1}$$

where we write only the significant digits of the estimate. The magnitude of the error is communicated by rounding the reported value to the appropriate number of digits. For example, given $a = \alpha \pm \delta$ the value a

$$\alpha = 3.14159265\,\text{kg}\,, \qquad \delta = 0.00014672\,\text{kg}\,, \tag{2}$$

$$\text{will be shown as} \quad a = 3.14159(15)\,\text{kg}\,, \tag{3}$$

$$\text{as opposed to} \quad a = (3.14159 \pm 0.00015)\,\text{kg}\,. \tag{4}$$

This notation is shorter, especially for very precise measurements and is very convenient in tables.

Listing 1: GNU OCTAVE code for the transformation of a value and its uncertainty into concise error notation

```
1   function pwe(val,dval)
    % Usage: pwe(val,dval)
    %   print with error
    %   converts a pair of value val and its
    %   uncertainty dval into concise notation
6   %   example: 1.2345 ± 0.0023 → 1.2345(23)
    s=floor(log10(val));  % magnitude of value
    ds=ceil(log10(dval)); % magnitude of uncertainty
    digits=s-ds+2;

11  fmt=sprintf("%%%i.%if(%%i)␣×␣10^{%i}\n",digits+1,digits,s);
    printf(fmt,val*10^(-s),round(dval*10^(2-ds)));
    endfunction
```

1. Motivation and Background

Sampling techniques can be used to fit models to available data, assist in experiment design, make prediction that include an uncertainty analysis, and select models based on evidence.

Various model classes exist within systems biology and typically the model states reflect the type of data that is available. The data can be static or dynamic, low or high resolution; ordinary differential equation models are compatible with all of these types. Of course, other model classes can be more suitable for a specific problem. But, the very general applicability of ODE models means that they can use data from many sources. The number of ODE models in systems biology is vast, more than 10^5 models are catalogued in the BioModels Database (Li et al. 2010).

Markov chain Monte Carlo is an increasingly popular method for parameter estimation with uncertainty analysis for such models. Currently available implementations of MCMC are limited to models with up to about 10^2 parameters. As computing hardware becomes faster, Bayesian techniques become applicable to increasingly large models. Current research includes models that challenge available MCMC algorithms. Models with about 30 to 50 parameters appear routinely in the literature (Dutta-Roy et al. 2015; Huang and Ferrel 1996; Hug et al. 2013), even larger models, with hundreds of parameters, are proposed increasingly often (Costa et al. 2014; Pokhilko, Mas, and Millar 2013). Faster MCMC algorithms will push the limit with regard to model sizes further upwards and alleviate the burden of parameter estimation for these very large models while still offering a reliable measure uncertainty to both fits and predictions. This is in contrast to optimisation methods, which for the most part forgo any uncertainty analysis.

Additionally, we observe that *steady state data* are persistently used to infer the interaction structure of reaction networks (Hecker et al. 2009; Khatri, Sirota, and Butte 2012; Vogel and Marcotte 2012) despite omiting the transient phase of the system dynamics. This type of data is comparatively cheap and is also available from high throughput measurement set-ups.

We exploit the steady state property of the data to do the calculations required for advanced MCMC methods a lot faster. We extend this approach to time series data and apply these new MCMC variants to models from the literature. The increased speed of these novel combinations of techniques makes it possible to tackle problems that might seem excessively large at first.

Another feature of data in biology is that it is often relative. Measurements obtained via (uncalibrated) densitometry of Western blots provide film *exposure levels* in arbitrary units rather than molecule numbers or substance concentrations. The quantification of Western blots is by no means an easy task (Gassmann et al. 2009; Taylor, Berkelman, et al. 2013; Taylor and Posch 2014). For the most part the measurements are normalised but not gauged. This type of data carries information when the results of an experiment are compared to a *control*, a neutral reference experiment. Western blots can also record a consistently scaled time series, measured to observe *trends*. In this setup, one of the points in the series is the reference point and is effectively eliminated to normalise the rest of the series. This normalising point doesn't need to be considered the neutral state of the system. For this reason we refer to both *controls* and normalising points more generally as *reference data*, which can be obtained in a *reference experiment*. Existing sampling tools lack support for relative data almost entirely and normalisation is circumvented by users through the introduction of additional unknown parameters into the measurement model. As this unnecessarily increases the problem size, we suggest the development of tools that support model output normalisation during sampling directly and provide such an implementation.

A more thorough description of modelling approaches in systems biology can be found in Appendix A, where we also expand on how systems biology relates to adjacent disciplines.

1.1. Contribution

This thesis provides a critical analysis of several numerical methods used in the Bayesian analysis of ODE models. We introduce a number of Markov chain Monte Carlo algorithms which provide a novel combination of existing techniques such as *analytical steady state sensitivity analysis*, the *Newton-Raphson method* and the application of the Fisher information as the metric tensor for transitions in parameter spaces.

Sampling procedures involve large numers of forward simulations of the model and can be very time consuming. Depending on the problem size, the used machine, and algorithm the sampling can last several hours or even days. Our goal is to reduce the sampling time and by doing so making the analysis of larger models feasible.

We have chosen advanced MCMC methods as our starting point. These algorithms are characterised by adaptation techniques using the model's local parameter sensitivity. We show that designing algorithms for specific types of biological data results in tremendous speedups.

We construct an SMMALA based algorithm that approximates the parameter sensitivity of the solution trajectories. The approximation is a lot easier to compute than the exact matrix and is most accurate for monotone convergence of the state trajectories from the initial conditions to stable steady states; it is exact for any steady state of the model. The new variant is called SSMMALA and works with time series data as well as with steady state data. We observe a speedup of one order of magnitude over the default implementation in our examples.

For steady state data, we have built two further variants of SMMALA: CBA+SMMALA and NR+SMMALA. The former uses the circuit breaking algorithm (Radde 2010) to directly calculate the steady states where possible, while the latter uses the Newton Raphson method and is more general. Both methods profit from fast, exact steady state sensitivity analysis, which informs the metric used during transitions. We observe an improvement of two orders of magnitude for CBA+SMMALA over SMMALA and one order of magnitude speedup of NR+SMMALA.

Additionally, we have made improvements for the RMHMC algorithm and achieved improvements up to three orders of magnitude for steady state data. In this manuscript we show why relative data can be problematic in numerical analysis and how it can be used with our algorithms.

In addition, we argue that the development of algorithms and software for scientific purposes in general can result in much more usable tools if more time and effort is invested for development. We have implemented some of the algorithms we discuss here as a C software as well as a script for high level tools such as MATLAB and GNU OCTAVE. Tests using problem sizes suitable for software development, with 10 to 20 parameters, immensely favour our C implementation, with speedups of *three* orders of magnitude.

For our evaluation, we use example models from literature as well as generic constructs or anonymised models from current research; we do not provide any additional insight into a *specific* biological problem. This approach keeps our implementation relevant for real world examples while keeping the focus of this work on the algorithms rather than the systems. We make no attempt to provide an overview of all MCMC algorithms but consider the ones that—to our knowledge—are common and have desirable properties such as ease of implementation, fast convergence, and good adaptation strategy.

In the following sections we provide a brief introduction into the notation of probability theory that will be used in this manuscript.

1.2. Probability

The variable names used here are unrelated to the rest of the manuscript. In some cases, we will calculate probabilities and probability densities of sets (e.g. the data $\mathcal{D}$), which contain random variables like the measurements and non-random parameters, e.g. standard deviations. We consider it understood that in such cases we integrate out the non-random parts.

As a general convention, we usually refer *by name* only to numerical values a random variable can take on. The random variable itself will remain unnamed outside of this section. We consider a random variable X to be a function that maps atomic random *events* $\omega \in \Omega$ to numerical values $x \in \mathcal{X}$ (say $\mathbb{R}^n_+$). Thus, we will deal with the arguments of probability density functions $p(x)$ (PDFS) and cumulative distribution functions (CDFS). Given a set of atomic *events* Ω and a sigma-algebra $\mathscr{F}$ on that set, the function $P : \mathscr{F} \to [0,1]$ will always signify the probability of a described event. The sigma-algebra is closed under Ω-complement operations, countable *unions and intersections* of its members; the power set of Ω is the largest sigma-algebra of Ω. So, P accepts logical combinations of atomic events.

A very common convention is to omit the event symbol ω and even the name of the random variable in arguments of P:

$$P(\{\omega : X(\omega) = x\}) =: P(X = x) \qquad \text{is abbreviated to } P(x)\,, \tag{1.1}$$

which we adhere to. In addition, expected values $\mathrm{E}(X)$ require the random variable's symbolic name, which we will abbreviate to $\langle x \rangle$, once again using the symbol for the random variable's values.

Probability distributions are specified by providing either the PDF or CDF (although other possibilities exist, e.g. characteristic functions). Probability distributions are to be understood as means of assigning a probability to random number values x falling into a given measurable set $\mathcal{A}$ (with measure $|\mathcal{A}|$). The probability of the value $x \in \mathcal{X}$ generated from a random process to lie within the measurable set A can be calculated using the PDF:

$$P(x \in \mathcal{A}) = \int_{\mathcal{A}} p(z)\,dz\,. \tag{1.2}$$

It follows, that a representative sample $\{x_k : k = 1, \ldots, N, x \sim p(x)\}$ from this distribution conforms to the expectation to find a relative frequency of values inside A in accordance with its probability mass (1.2). The expected value of a function of random variable is defined as follows:

$$\langle g(x) \rangle = \int g(z)p(z)\,dz\,. \tag{1.3}$$

A very useful *interpretation* of this value can be derived from (1.2). If we were to calculate the asymptotic average of a very large sample, we could construct a partition B_j of $\mathcal{X}$ ($\cup_j B_j = \mathcal{X}$) and make each B_j so small that $p(x)|_{x \in B_j} \approx$ const. Each B_j shall contain n_j sampled points and be represented by a characteristic point $b_j \in B_j$. We can predict n_j/N using the PDF. It follows:

$$\begin{aligned} [x] &= \frac{1}{N}\sum_{k=1}^{N} x_k \\ &\approx \frac{1}{N}\sum_j b_j n_j \\ &= \sum_j b_j \underbrace{\frac{n_j}{N}}_{P(x \in B_j)} \\ &= \sum_j b_j p(b_j)|B_j| \end{aligned} \tag{1.4}$$

Taking the limit of $N \to \infty$ asymptotically erases the difference $(x_k - b_j)$ between any x_k, which lies within a B_j and its representative centre b_j. This limit renders the approximation in (1.4) exact. This process converts the sum into an integral and consequently the measure $|B_j|$ into the integral's infinitesimal measure $dx_1 \ldots dx_m$. Thus, we obtain the equivalence relation:

$$\underbrace{\lim_{N \to \infty} \sum_j b_j p(b_j)|B_j|}_{\text{sample average}} = \int_X x p(x)\,dx_1 \ldots dx_m = \langle x \rangle\,. \tag{1.5}$$

For most standard probability distributions, $\langle x \rangle$ and other moments[1] can be calculated exactly in terms of the distribution's parameters. If that is not convenient or possible, the average of a sufficiently large sample can be used as an estimate of the expected value: $[g(x)] \approx \langle g(x) \rangle$. In summary, probability densities are tightly related to occurence frequencies: higher density values mean larger number of points in a given location.

[1] $\langle x^1 \rangle$ being the first moment

2. Introduction

Bayesian model analysis is an effective tool for the modelling framework of our choice: *stochastically embedded, intrinsically deterministic ordinary differential equation models*. It enables us to find model parameters that fit the noisy data in the most *probable* way for a given model. In the following Sections we give a short introduction into Bayesian model analysis. We put great emphasis on uncertainty analysis. Given uncertain data, we must include the propagated uncertainty into our predictions. To make this *goal* more tangible, we illustrate the desired outcome, an uncertain model fit, in the following Section without strict definitions or rigorous explanation. This is followed by a thorough look at the model structure, where we partition the model into *modules* such as *input*, *system*, and *output*. This view of models is omnipresent in systems theory and automatic control. Automatic control aims at constructing an input signal for the system to achieve a predetermined desirable behaviour in the broadest terms. This is not our goal. Yet, the *input/output* view is very useful. In contrast to engineering problems, we deal with systems that have unchangeable input dynamics. Therefore, we consider the input *dynamics* a part of the broader input model, while the *parametrisation* of these dynamics can be influenced during the experiment. One such example is an *exponential decay* of an administered drug, where only the *dose* can be chosen, i.e. the initial value.

In other words: the input dynamics are an intrinsic part of the model, while the input parameters are not and can be adjusted. For these reasons, we will refer to the input parameters themselves as *model inputs* or plainly *inputs*.

2.1. Available Data in Relation to Modelling Frameworks

Qualitative biological data is comparatively easy to obtain, no absolute quantification needs to be performed and RNA chips will provide the data in bulk. In light of this, it is tempting to use a qualitative modelling approach like time discrete Boolean networks, since they can incorporate this type of data very naturally.

However, we aim to ultimately make quantitative predictions. Data obtained from Western blots (Burnette 1981; Renart and Sandoval 1984) is readily available and is often gathered when the system is in its stable steady state. We have taken this opportunity to use the steady state property to tremendously decrease the computation time during parameter fits via Markov Chain Monte Carlo (MCMC) sampling methods. The MCMC technique will be explained in Section 3.1 and 3.2.

Steady states can be calculated more quickly than a full, time dependent solution trajectory of an initial value problem. Because of faster model response computations, a Bayesian ODE model approach can incorporate steady state data faster than time series data, while still being able to make fine grained, quantitative predictions. A Bayesian model fit might also make the need to gather more informative, dynamic data apparent when predictions are too uncertain. This can inform the next cycle of experiments, retaining the previously obtained information. In Bayesian analysis, this information is carried by the prior information within the next cycle.[1]

In this manuscript, our examples follow the paradigm of *experiments and controls*, but all listed methods can easily be generalised to *relative time series* progressions. The following sections will thoroughly define our modelling framework, but let us illustrate the goal and purpose of our approach first: a Bayesian model fit that contains uncertainty information.

Example

Without going into the depths of methodology, we present an example for a model with available relative data from literature. The data is depicted alongside our fit in Figure 2.1. We have constructed a modified model of the Erk phosphorylation in the MAPK signalling cascade from the supplement of Fritsche-Guenther et al. (2011):

$$\begin{aligned} \dot{x}_1 &= \rho_1 \frac{u}{1+u} - (2\rho_1 \frac{1}{1+u} + \rho_2) x_1 + (\rho_2 - \rho_1 \frac{1}{1+u}) x_2 \,, \\ \dot{x}_2 &= \rho_1 \frac{1}{1+u} x_1 - \rho_2 x_2 \,, \end{aligned} \tag{2.1}$$

[1] today's posterior is tomorrow's prior

with output function

$$y = \frac{\bar{x}_2(\rho, u)}{\bar{x}_2(\rho, u_0)}, \tag{2.2}$$
$$\text{with} \quad u_0 = 1\,.$$

Our modification introduces a kind of negative feedback between the state variables x_1 (pErk) and x_2 (ppErk) – once and twice phosphorylated Erk, respectively – to the input variable u, which models the action of Raf and MEK on Erk. The feedback is implicit and is not apparent in the model equations (see original publication). The twice phosphorylated Erk is measured up to a scaling constant and normalised using the control, characterised by $u_0 = 1$.

This two-state-variable model fits the data-points well and our Bayesian approach indeed provides not only a fit, but a statistical uncertainty analysis as well (shaded areas of the plot). In the following sections, we will introduce the full methodology of Bayesian parameter estimation and how it deals with ill-posed problems and relative data.

2.2. Modelling Using Ordinary Differential Equations

postulate: demand

In our specific setting we postulate that models consist of four (intertwined) parts: (i) the system model; (ii) the input model; (iii) the output model and (iv) the measurement error model.

(i) System Model. The state variables $x \in \mathbb{R}^n$ of the differential equation typically represent the concentrations of intracellular substances such as specific {m,t,r,sn,si}RNA, lipids, proteins or other acids. Qualitative experiments can provide the interaction structure of these substances. We can build a dynamic, parametrised ordinary differential equation proposal from these interactions. The parameters $\rho \in {\mathbb{R}_+}^m$ typically represent reaction kinetic coefficients; they are per definition positive. The state variable and parameter numbers n and m characterise the *problem size*. The parameters ρ are necessary to make predictions with a given model, therefore we will make a strong distinction between these parameters and those from other modules which are only necessary for the inference. The inner structure of this module is described in Section 2.2.2.

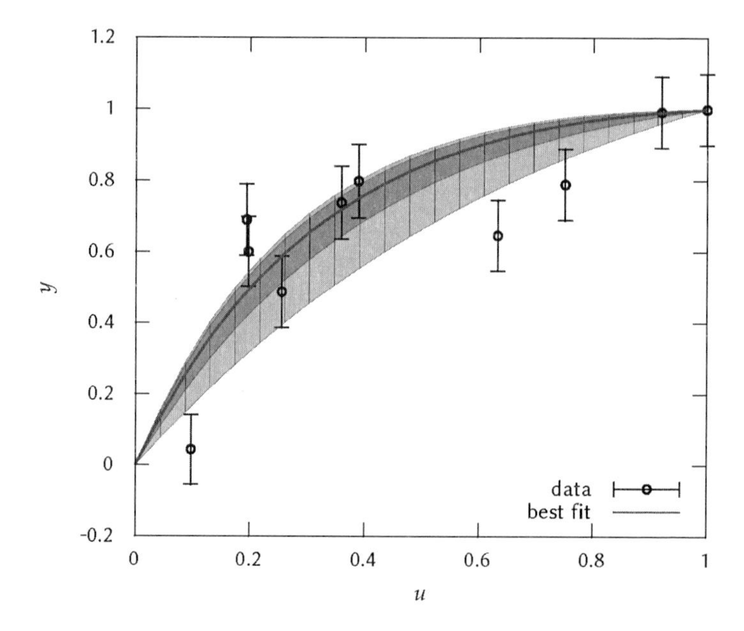

Figure 2.1.: Data from Fritsche-Guenther et al. (2011) and model prediction. We assumed a univariate, Gaussian noise model, with sensibly chosen parameters. The red (solid) line corresponds to a model fit, using the expected value of the parameters. The shaded patches represent an uncertainty analysis, the patches describe how the model response varies along the outskirts of the posterior distribution.

(ii) Input Model. Every experiment $\mathcal{E}$ influences this interaction structure in some way. The overall input may be a time dependent function within the model (e.g. a sigmoidal activation, or an exponential dissipation of an added substance). Since it is not obvious how exactly an experimental procedure changes the system, this experiment input function must usually be modelled as well; this module sometimes has its own unknown parameters. These are not needed when model assisted predictions for the system are made. We model the differences between experiments with different values for the inputs $u \in \mathbb{R}^{n_u}$ (constants), which can be influenced directly, i.e. set to some value within feasibility. If unknown input parameters are present, we estimate them alongside the intrinsic model parameters. The input functions themselves can be dynamic, but don't get a special symbol here. However, an input u_0 must exist which renders the input function neutral to model the control experiment (reference experiment).

Given initial conditions $x_0 = x(t_0)$ and parameters ρ, the model must have a solution trajectory, otherwise the parameters are discarded (deemed unlikely). Whenever no parameters can be found for which the initial value problem has a solution the model is discarded. The model can be checked for the existence and uniqueness of solutions by verifying Lipschitz continuity.

(iii) Output Model: An important part of the experiment's description is an output function $h(x, \rho, u)$, which models the current understanding of how the system's state, modelled by $x(t)$, relates to the observation $y \in \mathbb{R}^{n_y}$. While the input function is inseparably built into the model and necessary to make a forward simulation, the output is typically not fed back into the model and can be safely computed after the simulation. Nevertheless, some modelling languages (VFGEN, the SBPOP toolbox, SBML) allow the user to conveniently define the output function within the model.

(iv) Measurement Error Model. The measurements y are obtained with unavoidable measurement errors, the *noise*. Measurements reported without the parameters of a stochastic measurement error model cannot be interpreted, unless reasonable assumptions or guesses can be made. The stochastic description of the measurement process is the model we will use in Section 1.2 to construct a likelihood function for the data, and ultimately, to evaluate the parameters of the model in a

Bayesian sense.

2.2.1. Parameter Estimation

The unknown parameters ρ must be adjusted to reproduce observed data with a structurally known model. Many approaches exist and all of them have certain advantages and disadvantages. One increasingly popular method is the Bayesian approach coupled with Markov chain Monte Carlo sampling. It is particularly appealing for its relative simplicity and usefulness in cases of unidentifiable parameter vectors. The drawbacks lie in the high computational costs of MCMC methods.

The typical modelling problem, a good fit of observed data, is unfortunately often ill-posed in systems biology: many, hardly distinguishable[2] solutions exist. Naïve optimisation techniques might result in vastly different solutions depending on the starting point, all of which result in a good fit. A follow up question might be: *what is the extent of all well fitting parameters?* The prior probability distribution $p(\theta)$ can provide an effective regularisation to this problem (Radde and Offtermatt 2014).

The following section provides an introduction into the mathematical structure of the modelling modules listed in Section 2.2.

2.2.2. Model Structure Specifications in Relation to Data

The time dependent system states are captured by the state variables $x(t) \in \mathbb{R}^n$. The model parameters $\rho \in \mathbb{R}^m$ can describe the strength of interaction between the model state variables, or adjust the action of inputs on the model; most importantly, they are unknown. A second set of parameters $u \in \mathbb{R}^{n_u}$ describes the conditions of an experimental setup. These parameters can be set by the experimenter and may be external parameters, e.g. mechanical pressure on a tissue, but also describe internal modifications to the system, e.g. inhibitions to some of the interactions or the administration of a pharmaceutical compound. Each experiment $\mathcal{E}_k$ $(k = 0, 1, \dots, n_E)$ is defined by a specific input choice u_k and results in the observation of an experiment specific set of quantities, where $k = 0$ refers to the reference experiment. We collect these measurements and their uncertainties in the data set $\mathcal{D}$.

[2]Similar likelihood values

A given qualitative interaction structure and experimentally derived kinetics result in a parametrised ODE model:

$$\dot{x} = f(t, x; \rho, u_k)\,, \qquad k = 0, \dots, n_E\,, \tag{2.3}$$

$$x_{jk} = x(t_j; \rho, u_k)\,, \qquad j = 1, \dots, T\,, \tag{2.4}$$

$$x_{ijk} = x_i(t_j; \rho, u_k)\,, \qquad i = 1, \dots, n\,, \tag{2.5}$$

$$y_{jk} \stackrel{!}{=} h(x_{jk}, x_{j0}, \rho) + \Delta_{jk} + \varkappa_{jk}\,, \qquad \varkappa_{jk} \sim \mathcal{N}(0, \Sigma_{jk})\,, \tag{2.6}$$

where $y_{jk} \in \mathbb{R}^{n_y}$ is the output of the measurement process, which is recorded at T time points.[3] See Appendix C.1 on the handling of some special cases.

The exclamation mark in (2.6) indicates a modelling assumption. The values of y are taken directly from the data $\mathcal{D}$. The assumption is that the difference between these measurements and the model behaviour can in principle be explained by a random process $\varkappa_{jk}$ and a (small) systematic modelling error Δ_{jk}. The parameters Σ_{jk} of the random process should be estimated from the repetition of measurement, but the systematic error remains unknown. Most commonly, the noise components are independent from each other $\varkappa_{ijk} \sim \mathcal{N}(0, \sigma^2_{ijk})$ ($\Sigma_{jk} = \mathrm{diag}(\sigma^2_{\cdot jk})$). If the systematic error can be estimated, even partly, its influence on the data needs to be parametrised and included in the observation model h. The term Δ includes only ellusive, practically unknowable terms, by definition.

We assume that the noise $\varkappa_{jk}$ is the dominant contributor, $\Delta_{jk} \ll \sigma_{\cdot jk}$, (of $\varkappa_{jk}$) and perform the statistical analysis of the model without taking the systematic error into account. We will drop Δ and return to it, once we process model fits. A set of parametrised ODEs, with a specified input structure and an output function, are summarised as the overall model M.

Relative Data from Semi-Quantified Western Blots

Special considerations have to be made for the case of relative data, given in percent or arbitrary units as shown in Figure 2.2. The figure is taken directly from Möller et al. (2014); many more examples of relative data are shown in that publication. Typically, models will fit arbitrarily scaled data just as well as absolute data. However, the resulting parameters are useless for model comparisons with new, differently scaled data, since the old scaling is

[3]Whenever we omit arguments in h, we imply that there is no explicit dependence on these arguments in that specific case.

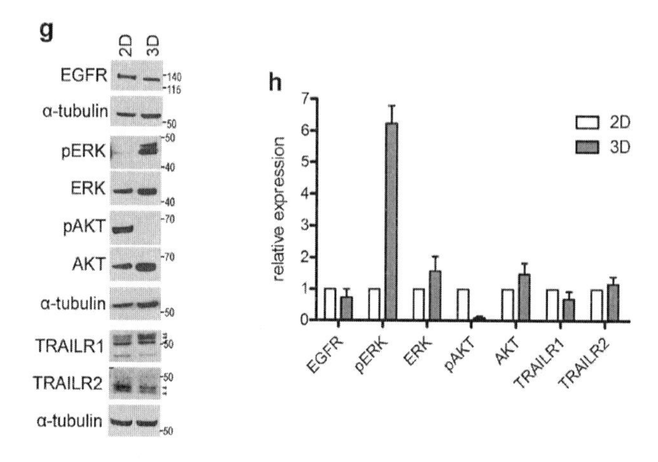

Figure 2.2.: Western blot data, quantified in relative terms. Published in Möller et al. (2014). (left) Raw Western blot with annotations. (right) Quantified values, relative to the control column labelled as 2D. In this case 7 reference values are provided by the control.

reproduced by the model. Therefore we have to ensure that reproducibility by normalising the data first.

This has implications for the output function h. Biologists typically take great care to obtain Western blots in the linear regime of the measurement equipment and adjust antibody concentrations and film exposure times accordingly. In such cases, the *primary* observation function g is linear in x

$$g(t,x,\rho,u) = Cx(t;\rho,u) \qquad h_i(t_j,x,\rho,u_k,u_0) = \frac{g_i(t_j,x,\rho,u_k)}{g_i(t_j,x,\rho,u_0)}, \tag{2.7}$$

but overall h forms a relation between an experiment and the control $k = 0$. The linear observations are characterised by the constant matrix C. As we have alluded to earlier we refer to (2.7) as a *two layer* output function; the first layer being the linear function g, the second layer being the ratio normalisation which mixes the two model simulations. The measurement time points t_j are not necessarily identical for $\mathcal{E}_{k\neq 0}$ and $\mathcal{E}_0$ in (2.7), but they are in all *examples* within this manuscript.

The real matrix $C \in \mathbb{R}^{n_y \times n}$ models the capabilities of the measurement setup. Specifically, the setup determines the sparsity and sign structure of C. But, the absolute value of the matrix elements remain very often unknown. A representative example for this linear observation model is

$$C = \begin{pmatrix} c_1 & 2c_1 & 0 \\ 0 & c_2 & -c_2 \end{pmatrix} = \begin{pmatrix} c_1 & 0 \\ 0 & c_2 \end{pmatrix} \begin{pmatrix} 1 & 2 & 0 \\ 0 & 1 & -1 \end{pmatrix} =: \operatorname{diag}(c) B\,. \tag{2.8}$$

Assuming that C is always scaled equally per row, with unknown row coefficients c_i, we can eliminate them[4] by taking row wise ratios between experiments. This is motivated by two properties of biological experiments: some molecule pairs cannot be separated within the chosen experimental setup and can only be quantified in sums; the unknown coefficient emerges during quantification and reflects conditions such as exposure times. The same quantification conditions apply to all simultaneosly measured targets. Known, constant factors (-1 and 2 in the above example) can be included, but are not typical in systems biology. In effect, y becomes a fully quantitative data matrix and does not depend on the unknown scaling parameters c anymore. Therefore, we are free to use the bare interaction structure B instead of C:

$$\begin{aligned} h_i(x_{jk}, x_{j0}, \rho) &= \frac{\sum_l C_{il} x_{ljk}}{\sum_l C_{il} x_{lj0}} \\ &= \frac{\sum_l c_i B_{il} x_{ljk}}{\sum_l c_i B_{il} x_{lj0}} \\ &= \frac{\sum_l B_{il} x_{lj0}}{\sum_l B_{il} x_{lj0}}\,. \end{aligned} \tag{2.9}$$

Sometimes, depending on modelling language (e.g. in VFGEN), an output function can be defined inside the model description file. The unavoidable consequence is that the output function can only refer to exactly one specific forward simulation of the model (a first layer function). This is why we split our notion of output model into two layers: a primary output function g of *exactly one* simulation, which returns a matrix $z_{ijk} = g_i(x(t_j), \rho, u_k) + \varkappa^g_{ijk}$, and a second layer which can perform ratios between simulations, e.g.: $y_{ijk} = z_{ijk} / z_{ij0}$.

[4]The unknown measurement coefficients c_i are uninteresting for the purpose of making predictions, since they change every time an experiment is performed.

The only remaining point to address is the standard deviation σ attached to y (the effective noise). Ratio-distributions are rarely simple elementary probability density functions. But, since we are interested in *uncertainty quantification* rather than rigorous statistics, we approximate the probability density of the ratio with another Gaussian density. The exact PDF of the ratio inherits some attributes from the primary noise Gaussians of $\varkappa^{g}$. Most importantly the ratio density is *unimodal*, which supports our approximation.

Given the standard deviations of the primary observables

$$z_{jk} = Cx_{jk} + \varkappa^{g}_{\cdot jk}, \qquad \varkappa^{g}_{ijk} \sim \mathcal{N}(0, \varsigma^{2}_{ijk}) \tag{2.10}$$

are known initially. The error propagates from z to y:

$$y_{ijk} = h_i(z_{jk}, z_{j,0}, u_k), \tag{2.11}$$

$$\sigma_{ijk} \approx \sum_{l} \left| \frac{\partial h_i(z_{jk}, z_{j0}, u_k)}{\partial z_{ljk}} \right| \varsigma_{ljk} + \sum_{l} \left| \frac{\partial h_i(z_{jk}, z_{j0}, u_k)}{\partial z_{lj0}} \right| \varsigma_{lj0}, \tag{2.12}$$

which is a linear approximation that typically overestimates the true variance. Of course, many different measurement error models exist and must be considered on a case to case basis; this is especially important, when the measurements are very precise. Other noise models are of course compatible with MCMC methods as well.

2.2.3. Graph Representation of ODE Models

A graph, consisting of *vertices* representing the interacting *substances* and *edges* between those vertices representing chemical reactions or more generally *interactions* between the compounds is a concise representation of a biological model. This representation is more abstract and can cover many model types such as Bayesian models, Boolean networks, and ODE models. In the latter case, the vertices are labelled consistently with the state variables while the edges represent fluxes of the ODEs right hand side. In biology, certain network motifs (subgraphs) tend to reappear in different organisms, these motifs are discussed in Alon (2007).

A directed edge e_{ji} between the vertices x_j and x_i is present if $\partial_{x_j} f_i(x, \rho, u)$ does not vanish:

$$(x_j) \rightarrow (x_i) \quad \Leftrightarrow \quad J_f(x, \rho, u)_i^{\,j} \neq 0$$

This representation of models is immensely useful and widely used in systems biology (along with other graphical representations). Graphical models can benefit from information gained from Bayesian analysis and can be used to inspect the MCMC sampling results (Vehlow et al. 2012).

The typical case for biological models is that graphs are directed but not free of cycles. Directed cycles are closed paths along the graph's edges, following edge direction, that contain two or more nodes. Degradation rates result in *loops*, an edge pointing to the vertex it originates from; loops are in this way different from cycles. The presence of directed cycles can make the analytical calculation of steady states harder. This graph structure is sufficient to describe all ODE models. Especially when they are constructed from the law of mass action, the graph structure is intuitive, since it shows conversion of substances. However, one often sees more elaborate graphs which more directly coincide with our intuition about chemical reactions involving *modifiers*. For example, the presence of enzymatic reactions requires more complex terms such as the Michaelis-Menten kinetics (Chen, Niepel, and Sorger 2010):

$$\mathrm{S} + \mathrm{E} \longleftrightarrow \mathrm{ES} \longrightarrow \mathrm{E} + \mathrm{P} \quad \Leftrightarrow \quad \dot{x}_{\mathrm{P}} = \theta_1 \frac{x_{\mathrm{S}} x_{\mathrm{E}}}{\theta_2 + x_{\mathrm{S}}} \tag{2.13}$$

In enzymatic reactions the mass action rule applies to the intermediate assembly of enzyme-substrate-complexes ES. These complexes are usually not part of the model. Instead, a summary of the overall reaction is modelled directly. This reduces the number of effective parameters and results in a multitude of useful kinetic laws for cases such as competitive inhibition and cooperative binding. The presence of an inhibitor or catalyst (modifier) is represented by edges that point from the modifier to another edge, specifically the edge between substrates and products. Enzymatic reactions make the use of more elaborate graphs tempting, since they not only visually represent the *role* of substances but also imply the use of special *enzyme kinetics* in the dynamical model.

Although these and many other classes of graphical models can be very useful, we will restrict our discussion to the simpler class of directed graphs $\mathcal{G}(\mathcal{V},\mathcal{E})$, where the edges in $\mathcal{E}$ are always arcs between vertex pairs from $\mathcal{V}$ and represent non-vanishing entries of the Jacobian J_f.

A directed, acyclic graph is abbreviated as DAG. In Section 5.1.3 we describe a method for the calculation of steady states in ODE models using graph theory related methods, which involve the transformation of a general regulatory network into an acyclic form.

Models in systems biology often require the Jacobian to retain a fixed sign structure. In the language of graphs, this is equivalent to the structure of $\mathcal{G}(\mathcal{V},\mathcal{E})$ remaining constant. This has to be enforced externally as incorrect parameters in the model can change edge direction. In systems biology, it is common to represent a dynamic model by a constant stoichiometry matrix R_S and a strictly positive flux vector $v(x,\rho,u)$

$$\dot{x} = R_S v(x,\rho,u)\,, \tag{2.14}$$

where each flux $v_l(x,\rho,u)$ is associated with the kinetic law of reaction l. The stoichiometry represents the conversion of substances in the model; it is fixed entirely and most often sparse. This form implies a fixed sparsity and sign structure of J_f and in turn non-negative parameters ρ. Consequently, it is safe to log-transform ρ to avoid negative values altogether.

2.2.4. Steady State Data

At least three types of special system behaviours can be reproduced by ODEs: oscillations, convergence to steady states and chaos. As far as we know, chaotic behaviour is not encountered in intracellular interactions. The convergence to steady states and oscillations are encountered frequently and have more mathermatical structure than transient model states. While we do not discuss oscillations in this manuscript, we do discuss the special properties of steady states and the benefits these properties entail for sampling algorithms.

It might be non-trivial to tell whether a system is in its steady state, i.e. if it is time invariant. Ultimately, this should be decided by the person who performs the experiment. We assume that the time scale for transient effects $\tilde{t}$ is known and successive measurements have been performed to ensure the time invariance of suspected steady states. We denote such steady states by omitting a time index and mark x with a bar:

$$\bar{x}_k \in \mathbb{R}^n : \quad 0 \stackrel{!}{=} f(\bar{x}_k(\rho,u_k),\rho,u_k)\,. \tag{2.15}$$

In some cases, (2.15) has several solutions. In such cases, we shall enumerate them using an additional index in braces: $\bar{x}_k^{\{1\}},\bar{x}_k^{\{2\}},\dots,\bar{x}_k^{\{r\}}$. Note, that all of these symbols represent vectors. Next, we discuss the sensitivity of states with regard to parameter changes.

2.2.5. Model Output Sensitivity

The sensitivity of $x(t;\rho,u_k)$ to parameter changes (ρ) is derived from f:

$$S_f(t,\rho,u_k)_i^j = \frac{\partial x_i(t,\rho,u_k)}{\partial \rho_j}\,. \tag{2.16}$$

One method of obtaining S_f in practice is to use a numerical ODE solver which performs sensitivity analysis during integration.

The output sensitivity S_h can be calculated easily from the state sensitivities (chain rule). As an example, we consider the simplest case covered by (2.7) with only one nonzero element c_i in each row of C, identified by the column index $I(i)$:

$$\begin{aligned} C_{is} &= \begin{cases} c_i & s = I(i) \\ 0 & \text{otherwise} \end{cases}, \\ y_{ijk} &= \frac{x_{I(i)jk}}{x_{I(i)j0}}\,. \end{aligned} \tag{2.17}$$

The sensitivity of h in terms of the presumably known state sensitivities $S_f(t;\theta,u_k)$ then is:

$$S_h(t_j;\rho,u_k)_i^l = \frac{\partial}{\partial \rho_l}\frac{x_{rjk}}{x_{rj0}}, \qquad \text{with } r = I(i) \tag{2.18}$$

$$= \frac{S_f(t_j;\rho,u_k)_r^l x_{rj0} - x_{rjk} S_f(t_j;\rho,u_0)_r^l}{{x_{rj0}}^2}$$

$$= \frac{S_f(t;\rho,u_k)_r^l - y_{rjk} S_f(t_j;\rho,u_0)_r^l}{x_{rj0}}\,. \tag{2.19}$$

Sensitivities will become very useful for the calculations of posterior density gradients in parameter space. They are also intimately related to the *natural metric* of the probabilistic parameter space which will be discussed in Section 3.3.

2.2.6. Logarithmic Space

Since the ODE models we consider in this manuscript have a fixed sign structure and consequently *non-negative parameters* we are free to operate in logarithmic space. In some cases models of biochemical reactions become unstable after certain sign flips and diverge, which makes an explicit restriction to one orthant prudent.

Sampling in logarithmic space θ and passing $\rho = \exp(\theta)$ to the ODE system has the additional benefit of covering several orders of magnitude more efficiently. This choice implies modifications to the output sensitivities.

$$\begin{aligned} S_h(t;\rho,u_k)_i^j &= \partial_{\rho_j} h_i(x(t;\rho,u_k),x(t;\rho,u_0))\,, \\ \partial_{\theta_j} h_i(x(t;\rho,u_k),x(t;\rho,u_0)) &= \sum_l \frac{\partial h_i(x(t;\rho,u_k),x(t;\rho,u_0))}{\partial \rho_l}\frac{\partial \rho_l}{\partial \theta_j} \\ &= S_h(t;\rho,u_k)_i^j \rho_j\,. \end{aligned} \tag{2.20}$$

and consequently, for any input w:

$$\begin{aligned} \partial_{\theta_k} S(t;\rho,w)_i^j \rho_j &= \sum_l \frac{\partial S_h(t;\rho,w)_i^j}{\partial \rho_l}\frac{\partial \rho_l}{\partial \theta_k}\rho_j + S_h(t;\rho,w)_i^j \frac{\partial \rho_j}{\partial \theta_k} \\ &= \frac{\partial S_h(t;\rho,w)_i^j}{\partial \rho_k}\rho_k\rho_j + S_h(t;\rho,w)_i^j \rho_k \delta_{jk}\,. \end{aligned} \tag{2.21}$$

We can now freely transform between the logarithmic sampling space and the given model structure in default space.

2.3. Measurement and Prediction Uncertainty

Straightforward methods exist to build an ODE model hypothesis from appropriate experiments[5]. Once the structure of the model is known the free *parameters* of the model need to be *estimated* using available data. In many cases, certainly for biological systems, measurements of the system are time consuming and expensive and in practice observations remain incomplete, such that the state of the system *cannot* be inferred algebraically, e.g. by means of an *inverse function* h^{-1}; the states are said to be unobservable. Conditions for observability and the design of observers that infer states from outputs are vast fields of study with many results for linear and nonlinear systems (Kalman 1970; Kazantzis and Kravaris 1998; Rami, Cheng, and Prada 2008).

[5] The experiments themselves can, of course, be very complex and quite expensive.

In some cases, the model parameters are structurally unidentifiable. A comparison of methods for identifiability analysis can be found in Chis, Banga, and Balsa-Canto (2011). For these reasons, but also due to limited amount of data points, the search for parameter vectors that fit the observations well can be an *ill-posed* inverse problem.

If the data is measured with large systematic and/or statistical errors it is very desirable to propagate this uncertainty in data to the parameter space and from there into predictions. An unbroken chain of uncertainty analysis is necessary for meaningful conclusions.

All of our examples will use the Gaussian error model without correlation between different measurements. Other error models are certainly possible as well; for example the log-normal error model is very sensible for strictly positive measurements. Correlations between measurements are also perfectly compatible with our approach and introduce no fundamental complications (except for a slight difference in notation).

This very simple approach to measurement errors proved to be very reliable with respect to omitted standard deviations in published data (see Appendix C.2).

2.3.1. Bayesian Model Analysis

The Bayesian method of model analysis attaches a probability density function to the parameter vector $\theta \in \mathbb{R}^m$. This density represents the state of knowledge or confidence in certain values. This distribution depends heavily on the data $\mathcal{D}$ and the model M. It is neither inherent to the true system nor does it imply stochasticity of the internal system but provides a measure of uncertainty in knowledge. The Bayes theorem

$$p(\theta|\mathcal{D},M)p(\mathcal{D}|M) = p(\mathcal{D}|\theta,M)p(\theta|M)\,, \tag{2.22}$$

applied to our framework, relates this posterior parameter distribution $p(\theta|\mathcal{D},M)$ to the likelihood function $L(\theta;\mathcal{D},M) = p(\mathcal{D}|\theta,M)$ and the prior distribution $p(\theta|M)$. The terminology regarding *likelihood* emphasises θ as a primary argument of L. L is not normalised, i.e. it doesn't integrate to 1, nor even necessarily normalisable with respect to θ, it is however normalised as a probability density of $\mathcal{D}$.

For the error model described in Section 2.2.2

$$y_{ijk} - h_i(x_{jk},\theta,u_k) = \varkappa_{ijk} \sim \mathcal{N}(0,\sigma_{ijk})\,, \tag{2.23}$$

n_E the number of experiments

the likelihood function reads

$$L(\theta; \mathcal{D}, M) = \frac{1}{\prod_{ijk} \sqrt{2\pi\sigma_{ijk}^2}} \exp\left(-\frac{1}{2} \sum_{ijk} \left(\frac{y_{ijk} - h_i(x_{jk}, \theta, u_k)}{\sigma_{ijk}} \right)^2 \right) . \quad (2.24)$$

The resulting log-posterior is given by

$$\begin{aligned} \log(p(\theta|\mathcal{D}, M)) &= -\frac{1}{2} \sum_{ijk} \left(\frac{y_{ijk} - h_i(x_{jk}, \theta, u_k)}{\sigma_{ijk}} \right)^2 - \frac{1}{2} \sum_{ijk} \log\left(2\pi\sigma_{ijk}^2\right) \\ &\quad + \log(p(\theta|M)) - \log(p(\mathcal{D}|M)) \quad (2.25) \\ &= -\frac{1}{2} \sum_{ijk} \left(\frac{y_{ijk} - h_i(x_{jk}, \theta, u_k)}{\sigma_{ijk}} \right)^2 - nTn_E \log(2\pi) \\ &\quad - \sum_{ijk} \log(\sigma_{ijk}) + \log(p(\theta|M)) - \log(p(\mathcal{D}|M)) , \quad (2.26) \end{aligned}$$

$A\backslash b$ solves $Ax = b$ for x

where we see some terms that are independent of θ and cannot have any influence on optimisation, including sampling techniques. The evidence $p(\mathcal{D}|M)$ is difficult to estimate and also independent of θ. We are free to remove any constant parts of (2.26) and define:

$$\ell(\theta; \mathcal{D}, M) = -\frac{1}{2} \sum_{ijk} \left(\frac{y_{ijk} - h_i(x_{jk}, \theta, u_k)}{\sigma_{ijk}} \right)^2 - \frac{1}{2} \left\| \Xi \backslash (\theta - \mu) \right\|^2 , \quad (2.27)$$

using a Gaussian prior $\mathcal{N}(\mu, \Xi)$. The function ℓ is proportional to $\log p(\theta|\mathcal{D}, M)$ and therefore equivalent for sampling purposes.

The prior distribution $p(\theta|M)$ is a natural way to introduce regularisation into the estimation problem. The parameter sample itself can be used for forward model simulations to obtain predictive distributions in the ODE state variables x.

The method of Bayesian model analysis uses random numbers drawn from the target distribution (2.27). While standard probability distributions, such as the uniform, Gaussian, Gamma, or Beta distribution are covered by dedicated random number generators, Markov Chain Monte Carlo (MCMC) methods are more general numerical algorithms which can draw samples from *any* distribution. Standard random number generators usually return independent, identically distributed random numbers. The Markov chain members are generated in succession; they are not independent from each other, which is a drawback. Expected values of random processes estimated using auto-correlated samples of size N are less precise than estimates performed with the same amount of independent random numbers. The auto-correlation between successive Markov chain members is a measure of this interdependence. The following chapter offers more detail on this topic.

Many MCMC algorithms are variations of the Metropolis algorithm (W. K. Hastings 1970; Metropolis et al. 1953) as it is very elegant and easy to implement. These algorithms are typically isotropic and suggest random steps in all directions with the same frequency. This certainly becomes a problem when high correlations between parameters are present within the posterior distribution. Some directions may be very prohibitive due to low likelihoods while others are permissive due to low sensitivity of the model output (with respect to this particular parametric direction). Section 3.5 is dedicated to this type of algorithm.

Many MCMC algorithm variants and modifications address the problem of strong correlations. For example: the adaptive Metropolis algorithm (Haario, Laine, et al. 2006) estimates the covariance of the distribution during sampling and takes it into account when making its random update steps. This is done to make the moves more efficient to cover the target distribution in fewer steps.

We chose to implement the simplified manifold MALA (SMMALA) and RMHMC algorithms, two adaptive MCMC methods, which are described in Girolami and Calderhead (2011). We compare several different implementations of these algorithms for ODE models in various programming languages with respect to their efficiency. All mentioned algorithms are explained in the next chapter.

2.3.2. Goodness of Fit

As mentioned in Section 2.2.2, we neglect the systematic model error Δ for the statistical analysis. With hindsight, we can check how well the difference between model fit and data is explained by normal noise, once a model fit, represented by $\hat{\theta}$, is obtained (to the best of our ability).

For the case of normal noise, we consider the log-likelihood function $\log(L(\theta;\mathcal{D},M))$, excluding all constant terms. We evaluate the log Likelihood for the best fit $\hat{\lambda}$ using the χ_k^2 distribution[6], which assigns probability values

[6] The χ_k^2 distribution is identical to the Gamma(α,β) distribution with $\alpha = k$ and $\beta = 1$

to the sum-of-squares of k normal random numbers, like in our error model of choice. Using the CDF of the χ^2 distribution (for k degrees of freedom) we can calculate the probability to obtain any log-likelihood value equal to $\hat{\lambda}$ or better (lower χ^2):

$$\hat{\lambda} = -\frac{1}{2}\sum_{ijk}\left(\frac{y_{ijk} - h_i(x(t_j;\hat{\rho},u_k))}{\sigma_{ijk}}\right)^2, \tag{2.28}$$

$$\gamma(k,s) = \int_0^s r^{k-1}e^{-r}dr, \tag{2.29}$$

$$\Gamma(k) = \int_0^\infty r^{k-1}e^{-r}dr, \tag{2.30}$$

$$s_\mathrm{r} = \sum_{j=1}^{k}\left(\frac{\varkappa_j - \mu_j}{\sigma_j}\right)^2, \qquad \varkappa_j \sim \mathcal{N}(\mu_j,\sigma_j^2), \tag{2.31}$$

$$\chi_k^2 \text{ CDF} \quad P(s_\mathrm{r} < s) = \frac{\gamma\left(\frac{k}{2};\frac{s}{2}\right)}{\Gamma\left(\frac{k}{2}\right)}, \tag{2.32}$$

where γ is known as the lower incomplete Γ-function. With $s = 2|\hat{\lambda}|$ and the number of degrees of freedom $k = n_E T n_y$

$$P(s_\mathrm{r} < 2|\hat{\lambda}|) = \frac{\gamma\left(\frac{k}{2};|\hat{\lambda}|\right)}{\Gamma\left(\frac{k}{2}\right)}. \tag{2.33}$$

This relation yields a probability, which is easy to interpret. It quantifies the plausibility of the data being adequately explained by the model M with normal noise $\varkappa$.

Throughout this manuscript, we try to avoid the notion of a *best fit* and instead use the whole *range* of solutions that MCMC offers to consistently transfer the measure of uncertainty from data to prediction. On this occasion, the best fit offers a good consistency check for our framework, model and data. Of course, many other methods to determine the goodness of fit do exist. The advantage of this approach lies in its simplicity.

3. Existing Methods for Parameter Estimation via Markov Chain Monte Carlo

In this Chapter, we recall the most necessary methods used for the Bayesian analysis of ODE model parameters: *Monte Carlo* (Anderson 1986; Hammersley and Handscomb 1964), a method which uses random numbers to estimate deterministic quantities, and *Markov chains* (Markov 1906), a specific type of random process which facilitates the creation of random numbers from non-standard probability distributions. We continue with a look on the metric of parameter spaces, specifically when dealing with ODE models. This chapter concludes with a description of *algorithm families* for Markov chain Monte Carlo (MCMC) sampling.

3.1. Monte Carlo

A very simple example of this concept is the numerical estimation of any given (complex) volume or area. MC makes it possible to write a quick function that estimates any such area characterised by, e.g. the overlap of various geometric shapes. Given a set of geometric shapes (polygons, circles) in the form of indicator functions[1] $I_j(x)$, one for each shape j, we can calculate the overlap A area

$$A = \int \prod_j I_j(x) dx_1 dx_2 \qquad x \in \mathbb{R}^2 . \tag{3.1}$$

This quantity can be approximated numerically by performing an MC estimate $[A]$ in a rectangular hull $\mathcal{A} = \{(a,b) | a \in (a_1, a_2), b \in (b_1, b_2)\}$ around these shapes. Using N random numbers $\xi \in \mathcal{A}$ ($\xi_{k,1} \sim \mathcal{U}(a_1, a_2)$, $\xi_{k,2} \sim \mathcal{U}(b_1, b_2)$), we count the number of random points inside the overlap

$$\langle A \rangle \approx [A] = \frac{(b_2 - b_1)(a_2 - a_1)}{N} \sum_k \prod_j I_j(\xi_k) . \tag{3.2}$$

[1] in OCTAVE: `inpolygon()`

An example for this calculation is depicted in Figure 3.1, where a circle, a triangle, and a square overlap. Random points were drawn from the square (with unit area $|\mathcal{A}| = (b_2 - b_1)(a_2 - a_1) = 1$) as support. We checked the indicator functions to determine whether they lie inside all given shapes. Dots inside this overlap are depicted in black, they are *hits*; outliers are rejected and omitted from the plot. The ratio of *hits* to the number of points used (N) is proportional to the area of the overlap normalised by the random number support. Consequently, the right hand side of (3.2) can be understood very intuitively:

$$\frac{A}{|\mathcal{A}|} = \frac{n}{N}, \tag{3.3}$$

where n is the number of hits, i.e.

$$n = \sum_k \prod_j I_j(\xi_k)\,. \tag{3.4}$$

Of course (3.2) also fits into the formalism of expected values, using (1.5) from Section 1.2:

$$\begin{aligned}
A &= \int \prod_j I_j(x)\, dx_1 dx_2 \\
&= \int \prod_j I_j(x) p(x)^{-1} p(x)\, dx_1 dx_2 \\
&= \int_{\mathcal{A}} \prod_j I_j(x) \underbrace{(b_2 - b_1)(a_2 - a_1)}_{|\mathcal{A}|} \underbrace{(b_2 - b_1)^{-1}(a_2 - a_1)^{-1}}_{p(x)} dx_1 dx_2 \\
&= |\mathcal{A}| \int_{\mathcal{A}} \prod_j I_j(x)(b_2 - b_1)^{-1}(a_2 - a_1)^{-1}\, dx_1 dx_2 \\
&= |\mathcal{A}| \left\langle \prod_j I_j(\xi) \right\rangle \\
&\approx |\mathcal{A}| \left[\prod_j I_j(\xi) \right],
\end{aligned} \tag{3.5}$$

which is identical to (3.2). The more random numbers are used, the better our estimate of the area A becomes.

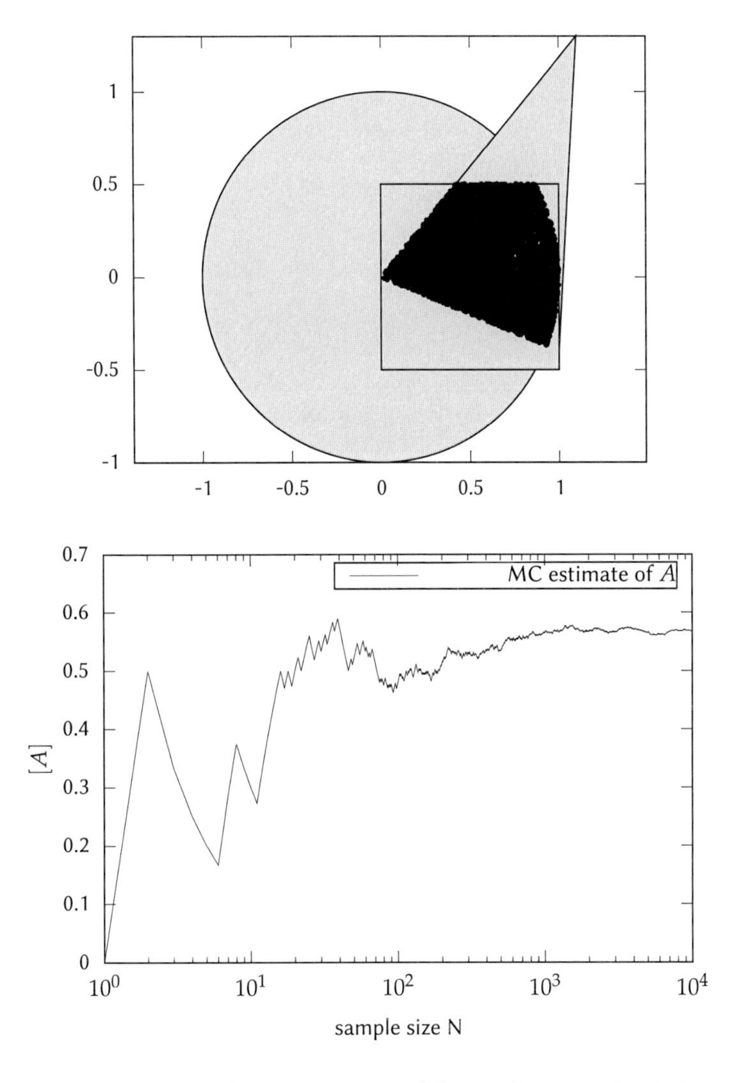

Figure 3.1.: Example – MC estimate of the overlap area of a circle, a square and a triangle: $[A] = 0.5723(49)$. We have used the square as the support for the uniform random number generator.

3.2. Markov Chains in the Reals

In all algorithms discussed in this manuscript, Markov chains are used to generate representative samples from probability distributions. Since we are interested in the parameter vectors θ, the chain members are real valued. The concepts introduced here are explained more thoroughly and discussed in Geyer (2011).

The shape of the target density is not an intrinsic property of the system that is being explored but represents the information content of the data for the used model. In the context of Bayesian learning the target distribution is called *posterior*.

A stochastic process producing values $\theta_i \in \mathbb{R}^m$, with discrete time index i, has the Markov property if the next step θ_{i+1} directly depends only on the current value θ_i and not on values prior to it. The stochastic transition rules $w : \theta_i \mapsto \theta_{i+1}$ imply a conditional transition probability density $p(\theta_{i+1}|\theta_i; w)$. The Markov property is best expressed in terms of these transition densities:

$$p(\theta_{i+1}|\theta_i, \ldots, \theta_0; w) = p(\theta_{i+1}|\theta_i; w)\,. \tag{3.6}$$

Despite θ_i being independent from any previous points except the direct predecessor θ_{i-1}, the members of the Markov chain may nevertheless be correlated over distances s:

$$\langle \theta_{i+s}\theta_i \rangle - \langle \theta_{i+s} \rangle \langle \theta_i \rangle =: \Gamma_\theta(s)\,. \tag{3.7}$$

Intense chain auto-correlation can be interpreted to mean that the process does not necessarily move very quickly away from previous values. We demand ergodicity from the Markov chains we use: given a target state space, the probability to get from any state θ to any other state ϕ must be non-zero. This may involve more than one iteration of the algorithm under scrutiny. Given a measurable subset Θ_1 in parameter space and an algorithm, represented by its characteristic transition probability density $p(\phi|\theta; w)$, we define an update step in terms of sets as $W : \Theta_1 \to \Theta_2$, with

$$P(\phi \in \Theta_2) = \int_{\Theta_1} p(\phi|\theta; w) P(\theta \in \Theta_1) d\theta\,. \tag{3.8}$$

Ergodicity demands this to be true for any two subsets of the state space:

$$\forall \Theta_1, \Theta_2 : P(W^s(\Theta_1) \cap \Theta_2) > 0\,, \tag{3.9}$$

for some (finite) integer s. For more information on ergodicity, especially how it applies to all algorithms described here, see Brooks, Gelman, et al. (2011) and Gelman, J. B. Carlin, et al. (2013).

Let us consider an ensemble of R such Markov Chains using the same update rules $w : \theta_{i,r} \to \theta_{i+1,r}$ $(r = 1, \ldots, R)$, then this ensemble at fixed times i represents a sample whose underlying density can be inferred by means of kernel density estimation. Applying the transition rules to each member will change the position of each chain and, if the process is not stationary, also the shape of the transient density. If the (inferred) density does not change during iteration, we call the process stationary. In this case the stationary distribution with rules w, $p(\theta|w)$, can be estimated just by observing one of the chains for a large number of update steps, since it will produce only points representative of this distribution.

For illustration, we have used the transition rules from Section 3.5.5[2] to generate $R = 128$ Markov chains; each starting randomly from the prior distribution (Gaussian). The resulting convergence process is depicted in Figure 3.2.

The stationarity condition in terms of transition probabilities is

$$\int p(\theta_a|\theta_b; w) p(\theta_b|w)\, d\theta_b = p(\theta_a|w)\,. \tag{3.10}$$

A sufficient condition for (3.10), called detailed balance, is beautifully symmetric:

$$p(\theta_a|\theta_b; w) p(\theta_b|w) = p(\theta_b|\theta_a; w) p(\theta_a|w)\,. \tag{3.11}$$

Integrating both sides of (3.11) with respect to θ_b yields (3.10):

$$\begin{aligned} \int p(\theta_a|\theta_b; w) p(\theta_b|w)\, d\theta_b &= \int p(\theta_b|\theta_a; w) p(\theta_a|w)\, d\theta_b\,, \\ \int p(\theta_a|\theta_b; w) p(\theta_b|w)\, d\theta_b &= p(\theta_a|w) \underbrace{\int p(\theta_b|\theta_a; w)\, d\theta_b}_{1}\,. \end{aligned} \tag{3.12}$$

Fortunately, if such a stationary distribution exists, it is unique and attracting for ergodic transition rules (Asmussen and Glynn 2011; Roberts and Rosenthal 2004). This means, that apart from numerical concerns, we can start the Markov chain anywhere (pick a point from any probability distribution within target's support) and let the imagined ensemble converge to the stationary distribution.

[2] arbitrarily selected

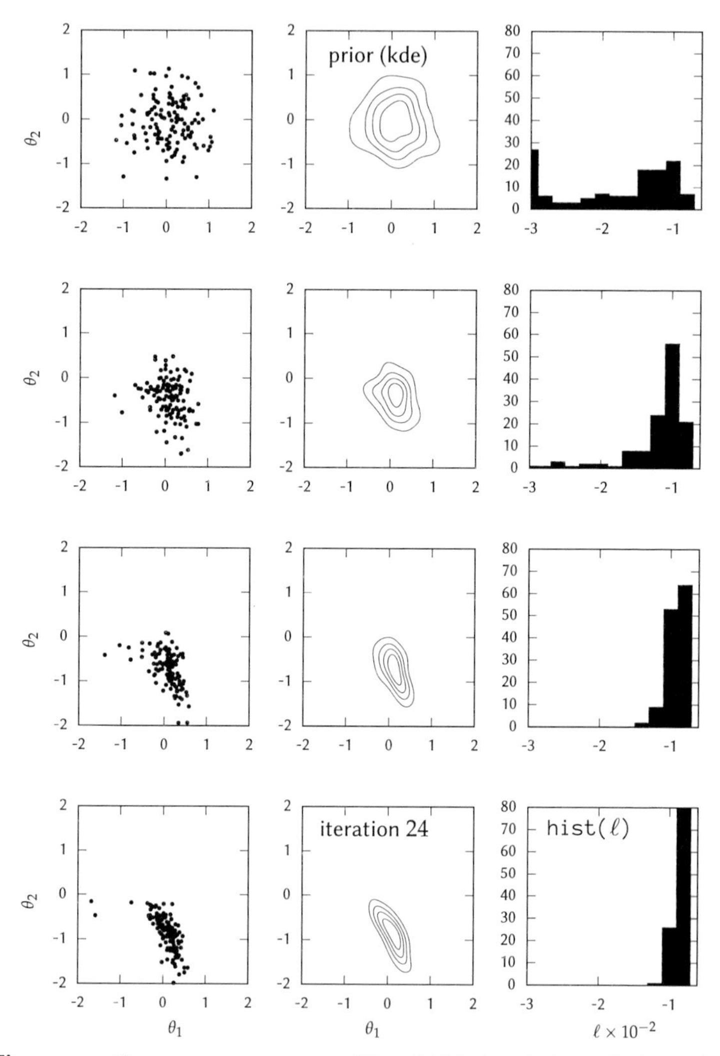

Figure 3.2.: The convergence process of $R = 128$ Markov chains with identical transition rules. The iteration index increases from top to bottom. The initial distribution is shown at the very top. *(left column)* Ensemble of chains. *(middle column)* kernel density estimate of the transient distribution. *(right column)* Histogram of the scaled log-posterior $\ell/100$; extreme values are lumped together at the left edge. The log-posterior values are spread over a huge area of lower probability at first, but progressively move towards higher values. Used model: Section 5.1.3 (5.7).

However, there are many practical concerns: it is impossible to tell whether all modes have been found. Most algorithms are sluggish when it comes to changing (or finding) separate modes. When a Markov chain initiates, it might be very far away from good values (large target density values). For unimodal distributions, only moves in a very specific direction (towards the mode) are acceptable. The rejection rate is very high, and consequently the convergence to the target is slow for this reason. Strategies such as thermodynamic cooling try to mediate the slow initial phase of MCMC.

Some parameter values or in fact wide regions in parameter space might be illegal arguments for the used model, i.e. the numerical method handling the model response does not return a valid model state. This increases the difficulty of finding the mode of the target distribution even more.

3.3. Geometry of Parameter Spaces

From the metric tensor, gradient and log-likelihood, we can define Monte Carlo algorithms which use Riemannian coordinates, rather than the standard Euclidean coordinate system for transitions in parameter space. Standard MCMC methods as well as their Riemannian manifold counterparts are listed in the next section. We refer the reader to Calderhead (2011) for further exposition of Riemannian geometry for MCMC. This article provides a comparison of traditional and Riemannian techniques. A classic result in statistics theory is the realisation that the Fisher information provides the proper distance metric for transitions between parameters of probability densities (Rao 1945, 1992; Skovgaard 1981).

We calculate the metric $G(\theta)^{rs}$ using the Fisher information of the likelihood function, but also the prior distribution distribution:

$$\begin{aligned} G(\theta)^{rs} &= \left\langle \frac{\partial \log p(\theta|\mathcal{D})}{\partial \theta_r} \frac{\partial \log p(\theta|\mathcal{D})}{\partial \theta_s} \right\rangle_{y|\theta} \\ &= \underbrace{\left\langle \frac{\partial L(\theta;\mathcal{D})}{\partial \theta_r} \frac{\partial L(\theta;\mathcal{D})}{\partial \theta_s} \right\rangle_{y|\theta}}_{G_y(\theta)^{rs}} - \underbrace{\left\langle \frac{\partial^2 \log p(\theta)}{\partial \theta_r \partial \theta_s} \right\rangle_{y|\theta}}_{-(\Xi^{-1})^{rs}}, \end{aligned} \tag{3.13}$$

where $G_y(\theta)$ is the expected Fisher Information of the likelihood and Ξ the covariance of the prior. We note that the normalising factor of the likelihood, and prior for that matter, is a constant with respect to θ and vanishes in the derivative of its logarithm. Fortunately, calculation of $G_y(\theta)$ requires only *first order* sensitivities $S_f(\rho,u)_l{}^r = \partial_{\theta_r}\bar{x}_l$. First order sensitivities of solution trajectories $x(t;\rho,u)$ are readily available from some ODE solvers such as the SUNDIALS suite.

For models with Gaussian measurement noise

$$
\begin{aligned}
G_y(\theta)^{rs} &= \left\langle \left(\frac{\partial}{\partial\theta_r}L(\theta;\mathcal{D})\right)\left(\frac{\partial}{\partial\theta_s}L(\theta;\mathcal{D})\right)\right\rangle ,\\
\text{with}\quad \frac{\partial}{\partial\theta_r}L(\theta;\mathcal{D}) &= \sum_{i_1 j_1}\left(\frac{y_{i_1 j_1}-h_{i_1}(\bar{x},\theta,u_{j_1})}{\sigma_{i_1 j_1}^2}\right) S_h(\bar{x},\theta,u_{j_1})_{i_1}^{\ r},\\
\text{and}\quad \frac{\partial}{\partial\theta_s}L(\theta;\mathcal{D}) &= \sum_{i_2 j_2}\left(\frac{y_{i_2 j_2}-h_{i_2}(\bar{x},\theta,u_{j_2})}{\sigma_{i_2 j_2}^2}\right) S_h(\bar{x},\theta,u_{j_2})_{i_2}^{\ s},\\
G_y(\theta)^{rs} &= \int\sum_{ij}\frac{(y_{ij}-h_i(\bar{x},\theta,u_j))^2}{\sigma_{ij}^2\sigma_{ij}^2} S_h(\bar{x},\theta,u_j)_i{}^r S_h(\bar{x},\theta,u_j)_i{}^s \prod_{\iota\kappa} p(y_{\iota\kappa}|\theta)dy_{\iota\kappa}\\
&= \sum_{ij} S_h(\bar{x},\theta,u)_i{}^r S_h(\bar{x},\theta,u_j)_i{}^s \int \frac{(y_{ij}-h_i(\bar{x},\theta,u_j))^2}{\sigma_{ij}^2\sigma_{ij}^2} p(y_{ij}|\theta)dy_{ij}\\
&= \sum_{ij} S_h(\bar{x},\theta,u_j)_i{}^r S_h(\bar{x},\theta,u_j)_i{}^s \frac{\sigma_{ij}^2}{\sigma_{ij}^2\sigma_{ij}^2}\,.
\end{aligned}
\tag{3.14}
$$

Taking the prior's contribution into account, this calculation yields the overall metric tensor $G(\theta)$, comprised of an inner product of the sensitivity matrices and the a-priori covariance matrix:

$$
G(\theta) = \sum_j S_h(\bar{x},\theta,u_j)^{\mathrm{T}}\Sigma_j^{-1} S_h(\bar{x},\theta,u_j) + \Xi^{-1}\,, \tag{3.15}
$$

where $\Sigma_j = \mathrm{diag}(\sigma_{\cdot j})$. $G_y(\theta)$ is the natural metric for the Riemannian manifold of Gaussians which form the likelihood function. From this metric the shortest distance trajectory, the *geodesic*, between the parameters of two samples of noisy model responses can be computed (Rao 1992). In other words, we can find the likelihood difference between two sets of parameters knowing only the parameters and the likelihood-metric $G_y(\theta)$. The second term Ξ^{-1} does the same for the influence of the prior. This reveals an intuitive interpretation of G_y, it is a local inverse covariance of the likelihood.

In summary, the overall matrix $G(\theta)$ provides us with a very efficient way to suggest moves through parameter space, which result in roughly the same changes of the posterior density in all directions. Algorithms, which make use of this Riemannian manifold view of the posterior density are RMHMC and SMMALA. In Section 3.5.5 we will describe in more detail how the transitions informed by the metric tensor $G(\theta)$ are used within the overall sampling algorithm.

3.4. MCMC Algorithm Type Overview

Here, we give a very brief overview of the following Sections, which describe the most common MCMC algorithms. We restrict ourselves to the bare essence of these algorithms and refer to the specific sections for details. We will consider three types of sampling algorithms:

Metropolis type the transition is stochastic, possibly informed by the posterior's covariance.

Langevin diffusion type a deterministic step along the gradient of ℓ is followed by a stochastic step.

Hamiltonian type the transition is a deterministic trajectory in parameter space with random initial conditions.

Each consists of a proposal and acceptance step, but only the proposal of the candidate ϕ is different between the three types. Figure 3.3 shows a schematic of these transition instructions; in addition we show the delayed rejection paradigm in Figure 3.4, which can be combined with any of the types, although it is almost exclusively used with Metropolis type algorithms (shown).

3.5. Metropolis-Hastings Type Algorithms

Here we discuss a class of algorithms that share two specific steps: (i) a successor point ϕ is picked randomly from the current point's surroundings and (ii) this candidate is rejected or accepted according to a probabilistic rule which guarantees detailed balance for the Markov chain. This class is heavily based on W. K. Hastings (1970) and Metropolis et al. (1953). Although the algorithms discussed here are more general, we will restrict our phrasing of their rules to the continuous, real case.

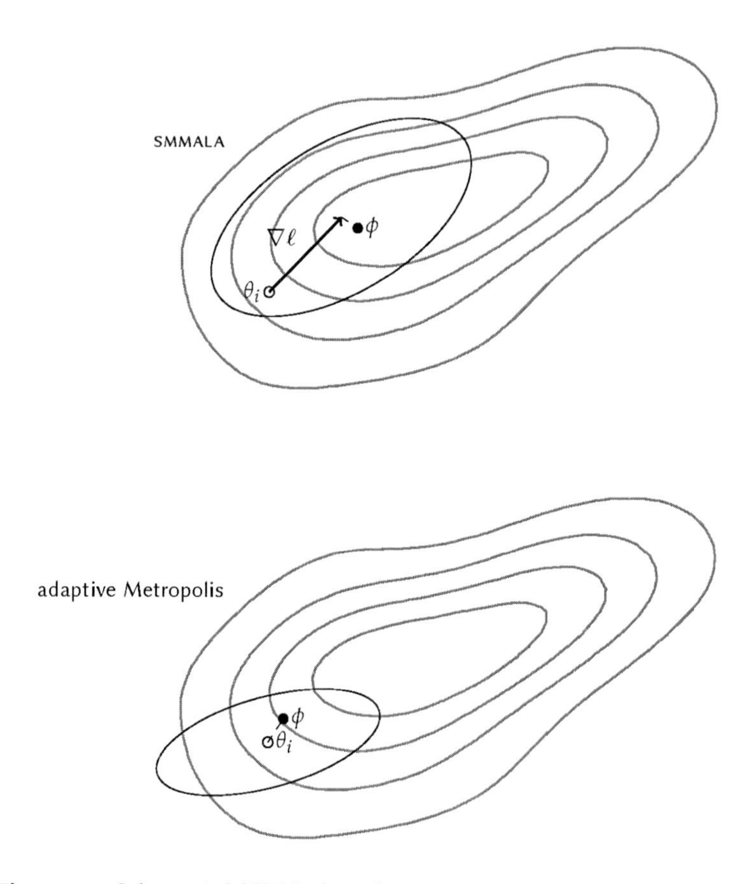

Figure 3.3.: Schematic MCMC algorithm type overview. Ellipses represent stochastic transitions using multivariate Gaussian distributions; filled dots represent the realisation of a stochastic move; arrows represent deterministic moves. Contour lines represent the target probability density. The hollow dot marks the current position, while ϕ labels the candidate point for the acceptance/rejection step of MCMC.

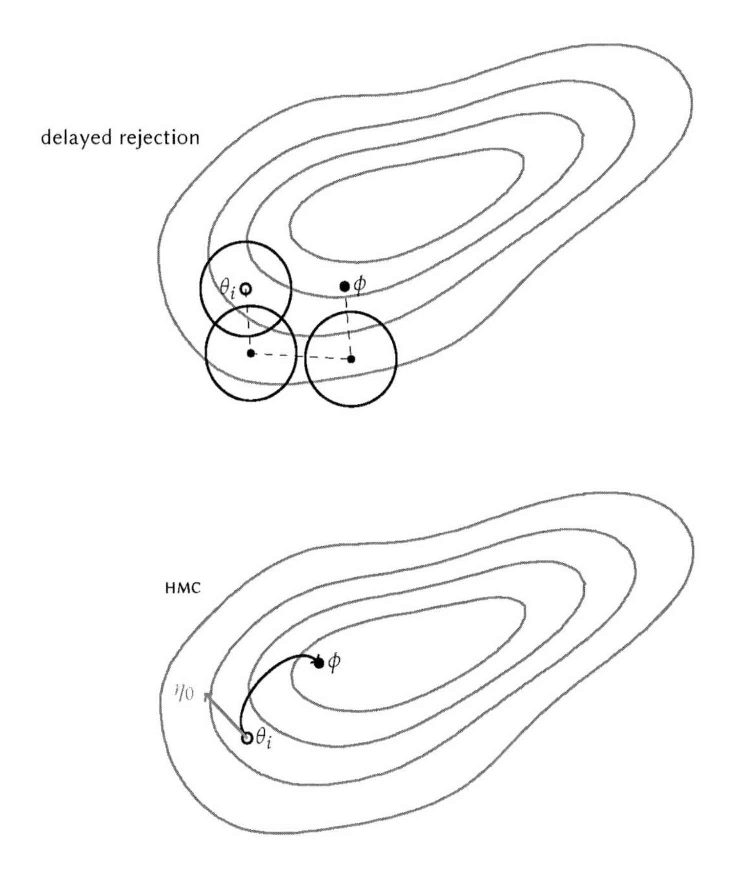

Figure 3.4.: Similar to Figure 3.3; in the case of delayed rejection, the intermediate points are rejected until either a maximum number of rejection is reached or acceptance occurs. As before: the hollow point marks the current position. In the case of HMC, an initial (random) momentum η_0 at this point is represented by a straight arrow.

Throughout this section we will usually use the symbol θ_i to represent the current point of the Markov chain. All members of the chain will be collected in the matrix $\Theta \in \mathbb{R}^{m \times N}$, where N is the sample size (component i of Markov chain member j is Θ_{ij}). As before an index-less θ represents symbolically the random process which generates the $\theta_i \in \mathbb{R}^m$ in expressions such as $\langle \theta \rangle$.

3.5.1. Metropolis

The successor ϕ is picked from a parametrised, symmetric probability density $p(\phi|\theta_i; b) = p(\theta_i|\phi, b)$, where b represents the proposition parameters (e.g. the covariance for Gaussians). The acceptance probability

$$\alpha(\phi; \theta_i) = \min\left(1, \frac{p(\phi|\mathcal{D}, M)}{p(\theta_i|\mathcal{D}, M)}\right) \tag{3.16}$$

ensures detailed balance (3.11):

$$p(\phi|\theta_i; b, \alpha) = p(\phi|\theta_i; b)\alpha(\phi|\theta_i) \tag{3.17}$$

$$p(\theta_i|\phi; b, \alpha) = p(\theta_i|\phi; b)\alpha(\theta_i|\phi) \tag{3.18}$$

with

$$\alpha(\phi; \theta_i) = \begin{cases} \frac{p(\phi|\mathcal{D},M)}{p(\theta_i|\mathcal{D},M)} & \text{if} \quad p(\phi|\mathcal{D}, M) < p(\theta_i|\mathcal{D}, M) \\ 1 & \text{else} \quad p(\phi|\mathcal{D}, M) > p(\theta_i|\mathcal{D}, M) \end{cases} \tag{3.19}$$

$$\alpha(\theta_i; \phi) = \begin{cases} 1 & \text{if} \quad p(\phi|\mathcal{D}, M) < p(\theta_i|\mathcal{D}, M) \\ \frac{p(\theta_i|\mathcal{D},M)}{p(\phi|\mathcal{D},M)} & \text{else} \quad p(\phi|\mathcal{D}, M) > p(\theta_i|\mathcal{D}, M) \end{cases} \tag{3.20}$$

in either case, we get:

$$\frac{p(\phi|\theta_i; b, \alpha)}{p(\theta_i|\phi; b, \alpha)} = \frac{p(\phi|\mathcal{D}, M)}{p(\theta_i|\mathcal{D}, M)}. \tag{3.21}$$

Comparing (3.21) with (3.11) reveals that the target distribution $p(\theta|\mathcal{D}, M)$ takes on the role of the stationary distribution, while α and b represent the rules of the transition:

$$p(\phi|\theta_i; b, \alpha)p(\theta_i|\mathcal{D}, M) = p(\theta_i|\phi; b, \alpha)p(\phi|\mathcal{D}, M). \tag{3.22}$$

Since the choice of the proposal density $p(\phi|\theta_i; b)$ is otherwise free, we may choose the normal distribution.

3.5.2. Metropolis-Hastings

An important variation of this algorithm allows asymmetric proposals, with $p(\phi|\theta_i;b) \neq p(\theta_i|\phi;b)$ (W. K. Hastings 1970). The acceptance probability needs an additional factor to compensate:

$$\alpha(\phi;\theta_i) = \min\left(1, \frac{p(\phi|\mathcal{D},M)}{p(\theta_i|\mathcal{D},M)}\frac{p(\theta_i|\phi;b)}{p(\phi|\theta_i;b)}\right) \tag{3.23}$$

which once again, in close analogy to (3.19) and (3.20), ensures detailed balance:

$$\begin{aligned}
\frac{p(\phi|\theta_i;b,\alpha)}{p(\theta_i|\phi;b,\alpha)} &= \frac{p(\phi|\theta_i;b)}{p(\theta_i|\phi;b)}\frac{\alpha(\phi;\theta_i)}{\alpha(\theta_i;\phi)} &(3.24)\\
&= \frac{p(\phi|\theta_i;b)}{p(\theta_i|\phi;b)}\frac{p(\phi|\mathcal{D},M)}{p(\theta_i|\mathcal{D},M)}\frac{p(\theta_i|\phi;b)}{p(\phi|\theta_i;b)} &(3.25)\\
&= \frac{p(\phi|\mathcal{D},M)}{p(\theta_i|\mathcal{D},M)}\,. &(3.26)
\end{aligned}$$

3.5.3. Adaptive Metropolis

In cases where proposed points $\phi \in \mathbb{R}^m$ are frequently rejected, i.e. the average acceptance rate is low, the Metropolis-Hastings algorithm becomes inefficient in estimating density properties. Unnecessarily large samples (and therefore Markov chains) have to be recorded to obtain an accurate description of the underlying stationary distribution. For such cases, the Metropolis algorithm can be modified to adapt to the target density's covariance structure (Haario, Saksman, and Tamminen 1998). This is done by choosing a multivariate Gaussian proposal density with an adapting covariance C_i^{am}, which progressively converges to the posterior covariance $\langle\theta\theta^T\rangle - \langle\theta\rangle\langle\theta\rangle^{\mathrm{T}}$, scaled down by the step size $s(m)$. To avoid frequent recalculations of the posterior covariance during sampling using the Markov chain history, a recursive update rule for the estimate can be used.

$v_{i<j}$ is similar to MATLABS v(1 : j−1)

The proposal, with diminishing adaptation, at Markov chain step i:

$$\vartheta_1 = [\theta; i-1] \,, \tag{3.27}$$

$$\vartheta_0 = [\theta; i] = \frac{(i-1)\vartheta_1 + \theta_i}{i} \,, \tag{3.28}$$

$$C^{\text{am}}_{i+1} = \frac{i-1}{i} C^{\text{am}}_i + \frac{s(m)}{i} \left(i\vartheta_1\vartheta_1^{\mathrm{T}} - (i+1)\vartheta_0\vartheta_0^{\mathrm{T}} + \theta_i\theta_i^{\mathrm{T}} + \epsilon I_m \right) \,, \tag{3.29}$$

$$\phi \sim \mathcal{N}(\theta_i, C^{\text{am}}_i) \tag{3.30}$$

where (3.29) is the recursive representation of the equivalent explicit definition $C^{\text{am}}_i = \text{cov}(\Theta_{\cdot j<i}) + s(m)\epsilon I_m / i$. A user defined scale parameter $s(m)$ can be used to tune the algorithm. Since C^{am}_i must be invertible; the additional parameter ϵ is used to mix in the diagonal unit matrix of size m (`diagones(1,m)`) to ensure this property. In practice, ϵ must be chosen large enough to overcome possible numerical problems with the invertibility of C^{am}_i or omitted if such problems do not occur.

An important limitation of this algorithm is the assumption that the posteriors global covariance represents the local shape of the density at the current point θ_i. For more complex distributions a local estimate of the posterior's curvature is more informative (see Section 5.3.3).

3.5.4. Delayed Rejection

The paradigm of delaying a rejection was introduced in Mira (2001). This technique aims at increasing the acceptance rate while simultaneously moving larger distances through parameter space, thus increasing the estimation accuracy of the Markov chain.

We have used this algorithm, since it is used frequently and performs quite well in many cases. However, since we do not perform an in depth comparison of this algorithm to the others, we describe only the basic idea and refer to the original publication and Haario, Laine, et al. (2006) for proofs of detailed balance and explanations of the rationale behind the rule construction. It is a rather intricate algorithm and not always easy to implement since it requires recursive acceptance rate calculation.

In contrast to a one step MCMC algorithm, a sequence of transition probability densities $q_l(\phi_l|\theta_i, \phi_{k<l})$ is used whenever a candidate point is

rejected. This implies a sequence of acceptance probabilities

$$\alpha_l(\phi_l; \cup_{j=1}^{l-1}\phi_j, \theta_i) = \min\left(1, \frac{p(\phi_l|\mathcal{D}, M)Q_l^{\mathrm{N}}(\theta_i; \cup_r\phi_r)A_l^{\mathrm{N}}(\cup_r\phi_r)}{p(\theta_i|\mathcal{D}, M)Q_l^{\mathrm{D}}(\cup_r\phi_r; \theta_i)A_l^{\mathrm{D}}(\cup_{r=1}^{l-1}\phi_r; \theta_i)}\right), \tag{3.31}$$

where

$$Q_l^{\mathrm{N}}(\theta_i; \cup_{r=1}^{l}\phi_r) = q_l(\theta_i| \cup_r \phi_r)\prod_{k=1}^{l-1} q_k(\phi_{l-k}| \cup_{r=l-k+1}^{l} \phi_r), \tag{3.32}$$

$$Q_l^{\mathrm{D}}(\cup_{r=1}^{l}\phi_r; \theta_i) = \prod_{k=1}^{l} q_k(\phi_k| \cup_{r=1}^{k-1} \phi_r, \theta_i), \tag{3.33}$$

$$A_l^{\mathrm{N}}(\cup_{r=1}^{l}\phi_r) = \prod_{j=1}^{l-1}(1 - \alpha_j(\phi_{l-j}; \cup_{r=l-j+1}^{l}\phi_r)) \tag{3.34}$$

$$A_l^{\mathrm{D}}(\cup_{r=1}^{l-1}\phi_r; \theta_i) = \prod_{j=1}^{l-1}(1 - \alpha_j(\phi_j; \cup_{r=1}^{j-1}\phi_r, \theta_i)). \tag{3.35}$$

Yet, delaying the rejection too often might end up to be very inefficient in terms of uncorrelated points per second of CPU time. Limiting the number of delays to about two might be of much practical use:

$$\alpha_1(\phi_1; \theta_i) = \min\left(1, \frac{p(\phi_1|\mathcal{D}, M)q_1(\theta_i|\phi_1)}{p(\theta_i|\mathcal{D}, M)q_1(\phi_1|\theta_i)}\right), \tag{3.36}$$

$$\alpha_2(\phi_2; \phi_1, \theta_i) = \min\left(1, \frac{p(\phi_2|\mathcal{D}, M)q_2(\theta_i|\phi_1, \phi_2)q_1(\phi_1|\phi_2)(1 - \alpha_1(\phi_1; \phi_2))}{p(\theta_i|\mathcal{D}, M)q_1(\phi_1|\theta_i)q_2(\phi_2|\phi_1, \theta_i)(1 - \alpha_1(\phi_1; \theta_i))}\right).$$

These rules can be used to construct Markov chains with better properties by applying them to different base algorithms. One very important example is the DRAM algorithm (Haario, Laine, et al. 2006), which combines delayed rejection with adaptive Metropolis. A matlab implementation of this algorithm has been released by the authors[3]. We consider this implementation a reference for algorithm testing and performance evaluation. We have also implemented a custom delayed rejection Metropolis algorithm (DRM) in GNU OCTAVE.

However, since this algorithm is most often combined with either Metropolis or its adaptive version, the result inherits some of the limitations of the latter.

[3] http://helios.fmi.fi/~lainema/mcmc/

3.5.5. Simplified Manifold Metropolis Adjusted Langevin Algorithm

The foundation for the SMMALA algorithm is discussed in Roberts and Tweedie (1996) while Girolami and Calderhead (2011) add the Riemannian manifold view to the algorithm.

For MALA, the Markov chain is a discrete time approximation of the stochastic differential equation (SDE)

$$d\theta = -\frac{1}{2}G^{-1}(\theta)\nabla_\theta V(\theta;\mathcal{D},M)dt + \mathrm{chol}(G^{-1}(\theta))db(t), \tag{3.37}$$

where $V(\theta;\mathcal{D},M) = -\log(p(\theta|\mathcal{D},M))$ and $db(t)$ is a Wiener process. The natural metric $G(\theta)$ of the parameter space is the Fisher information, *chol* denotes the Cholesky decomposition (returns one of the factors). The proposal $\theta_i \to \phi$ with step size $\epsilon \in \mathbb{R}_+$

$$p(\phi|\theta_i) = \mathcal{N}(\nu(\theta,\epsilon), C^{\mathrm{sm}}(\theta,\epsilon))\,, \tag{3.38}$$

$$\nu(\theta_i,\epsilon) = \theta_i - \frac{\epsilon}{2}G^{-1}(\theta_i)\nabla_\theta V(\theta;D)|_{\theta_i}\,, \tag{3.39}$$

$$C^{\mathrm{sm}}(\theta_i,\epsilon) = \epsilon^2 G^{-1}(\theta_i)\,, \tag{3.40}$$

takes information about the local shape of the posterior distribution into account. The metric tensor $G(\theta)$ is calculated using sensitivity analysis (S_h),

$$G(\theta_i) = \sum_{j,k} S_h(t_j;\theta_i,u_k)^T \Sigma_{j,k}^{-1} S_h(t_j;\theta_i,u_k)\,. \tag{3.41}$$

3.6. Hamiltonian Type Algorithms

The Hamiltonian Monte Carlo algorithm can be constructed by introducing an auxiliary variable to extend the state space. We may interpret the auxiliary variable as a momentum variable, and the negative log joint distribution may be interpreted as a Hamiltonian system (Girolami and Calderhead 2011; Radford 2011). The main idea is that the induced dynamics of this system may then be used for proposing moves through parameter space within an MCMC scheme. This is desirable since the dynamics may propose points that are far from the current point and accepted with high probability.

3.6.1. Hamiltonian Monte Carlo (HMC)

We begin by rewriting the posterior probability as

$$p(\theta|\mathcal{D},M) \propto \exp(-V(\theta;\mathcal{D},M)) \tag{3.42}$$

where,

$$V(\theta;\mathcal{D},M) = \frac{1}{2}\sum_{k=1}^{n_E}\sum_{j}^{T}\sum_{i=1}^{n_y}\left(\frac{y_{ijk} - h_i(x_{jk}, x_{j0}, \theta, u_k)}{\sigma_{ijk}}\right)^2 - \log(p(\theta))$$

for the measurement uncertainty modelled by (2.26) in Section 2.3. The sampling space is then extended by introducing the momentum variable η, and we may write the joint distribution as

$$p(\eta,\theta|\mathcal{D},M) = e^{-H(\eta,\theta)} = \exp\left(-\frac{1}{2}\eta^{\mathrm{T}}\eta\right)\exp\left(-V(\theta;\mathcal{D},M)\right). \tag{3.43}$$

We note that the Hamiltonian function $H(\eta,\theta)$ is simply the negative log joint distribution of the extended state space and can be used to calculate Hamiltonian trajectories according to the differential equations defined in the algorithm below. Given current values for the parameter and momentum variables, we can simulate the Hamiltonian dynamics to propose a new pair of parameter and momentum variables, which are then accepted according to a Metropolis-Hastings ratio to ensure convergence to the correct stationary distribution. The advantage of this approach is that this ratio may be close to 100 %, far higher than the typical optimal acceptance ratios for other MCMC algorithms, which are between 20 % and 60 %. The standard HMC algorithm is given by two steps:

1. **Transition step**
   Starting at $\theta =: \theta(0) = \theta_0$, solve the differential equations,

$$\begin{aligned} \partial_\tau \eta(\tau) &= -\nabla_\theta V(\theta(\tau);\mathcal{D},M), \\ \partial_\tau \theta(\tau) &= \eta(\tau), \end{aligned} \tag{3.44}$$

   for $\tau \in [0,\mathcal{T}]$ with initial conditions:

$$\begin{aligned} \eta(0) &\sim \frac{1}{\sqrt{2\pi}^m}\exp\left(-\frac{1}{2}\eta^T\eta\right), & \eta' &:= \eta(\mathcal{T}), \\ \theta(0) &= \theta_0, & \theta' &:= \theta(\mathcal{T}), \end{aligned} \tag{3.45}$$

where the proposed parameter and momentum variables at time $\mathcal{T}$ are given on the right. The above equations are Hamilton's equations of motion for a particle with momentum η in a potential field $V(\theta; \mathcal{D}, M)$.

2. **Acceptance step**
   Accept θ' and η' with probability

$$\begin{aligned}\alpha(\eta', \theta'; \eta, \theta) &= \min\left(1, \frac{p(\eta', \theta'|\mathcal{D})}{p(\eta, \theta|\mathcal{D})}\right) \\ &= \min\left(1, \exp\left(H(\eta, \theta) - H(\eta', \theta')\right)\right). \end{aligned} \tag{3.46}$$

The numerical solution of (3.44) must be calculated using a symplectic integrator (Girolami and Calderhead 2011), which is exactly time reversible and volume preserving; these properties are required for ensuring convergence to the correct stationary distribution. Since the Hamiltonian laws of motion conserve energy, an analytic solution would provide sample points with perfect acceptance. Numerical integrators introduce small integration errors and consequently the acceptance rate is slightly lower than 100%.

3.7. Riemannian Manifold Hamiltonian Monte Carlo

The RMHMC algorithm (Girolami and Calderhead 2011) is the most complex algorithm in the list of methods we discuss here. Like with SMMALA, the transition governing equations are extended by the introduction of a metric tensor:

$$H(\eta, \theta) = -\frac{1}{2}\eta^{\mathrm{T}} G(\theta)^{-1} \eta + \frac{1}{2} m \log(2\pi) + \mathrm{tr}(\log(\mathrm{chol}(G(\theta)))) + V(\theta). \tag{3.47}$$

The resulting Hamiltonian equations of motion are not separable. To guarantee reversibility an implicit integration method has to be used, the generalised leapfrog algorithm (Leimkuhler and Reich 2004).

This method requires an aditional iteration until a fixed point is reached for each update step. This increases the number of required simultaions for the model M dramatically. For this reason, the RMHMC algorithm stands to benefit much more than SMMALA from any improvement in model response calculations.

Although the convergence properties of this algorithm are generally improved over HMC the mechanism is very similar and the algorithms proceed identically after the integration is completed.

3.8. Sensitivity Analysis for Steady States

The calculation of the metric tensor $G(\theta)$ requires the sensitivities of the output function h (Section 3.3 and 2.2.5). Some numerical integrators, e.g. the SUNDIALS suite of solvers, perform a forward sensitivity analysis when calculating the solution for an initial value problem. In simulation frameworks that don't provide a sensitivity analysis, it is harder to implement algorithms such as SMMALA or RMHMC. In Chapter 5 we will discuss many variants on these two methods and also provide alternative methods for sensitivity analysis.

In the case of steady state data, the sensitivities can be obtained from the steady state condition:

$$\begin{aligned} 0 &\stackrel{!}{=} f_i(\bar{x}(\theta,w),\theta,w) \\ 0 &\stackrel{!}{=} \frac{df_i(\bar{x}(\theta,w),\theta,w)}{d\theta_j} \\ &= \sum_{l=1}^{n} \frac{\partial f_i(\bar{x},\theta,w)}{\partial \bar{x}_l} \frac{\partial \bar{x}_l(\theta,w)}{\partial \theta_j} + \frac{\partial f_i(\bar{x},\theta,w)}{\partial \theta_j}\,. \end{aligned} \tag{3.48}$$

We will drop the arguments of $\bar{x}$ to shorten notation. In (3.48) we see that the steady state sensitivity is obtained by solving the following linear algebraic equation,

$$\begin{aligned} 0 &= J_f(\bar{x},\theta,w)S_f(\theta,w) + K_f(\bar{x},\theta,w)\,, \\ \Rightarrow S_f(\bar{x},\theta,w) &= -J_f(\bar{x},\theta,w)^{-1}K_f(\bar{x},\theta,w)\,, \end{aligned} \tag{3.49}$$

where $K_f(\bar{x},\theta,w)_i^j = \partial_{\theta_j} f_i(\bar{x},\theta,w)$. We denote the solution to (3.49) as $S_f(\bar{x},\theta,w)$, which is easy to obtain when the Jacobian is invertible[4]. Similarly, we can write the following equation for the second order sensitivity (Girolami

[4]We note that this is not always the case: whenever conservation relations are present, for example, the Jacobian is not invertible anywhere. However, in such cases it is sufficient to use these conservation relations to reduce the number of state variables, as we do in the examples.

and Calderhead 2011),

$$
\begin{aligned}
0 = \frac{d}{d\theta_k}&\left(\sum_l \frac{\partial f_i(\bar{x},\theta,w)}{\partial \bar{x}_l}\frac{\partial \bar{x}_l(\theta,w)}{\partial\theta_j} + \frac{\partial f_i(\bar{x},\theta,w)}{\partial\theta_j}\right)\\
&= \sum_{r=1}^{n}\sum_{l=1}^{n}\left(\frac{\partial^2 f_i(\bar{x},\theta,w)}{\partial\bar{x}_r\partial\bar{x}_l}\frac{\partial\bar{x}_r}{\partial\theta_k} + \frac{\partial^2 f_i(\bar{x},\theta,w)}{\partial\theta_k\partial\bar{x}_l}\right)\frac{\partial\bar{x}_l(\theta,w)}{\partial\theta_j}\\
&\quad + \sum_{l=1}^{n}\frac{\partial f_i(\bar{x},\theta,w)}{\partial\bar{x}_l}\frac{\partial^2\bar{x}_l(\theta,w)}{\partial\theta_k\partial\theta_j}\\
&\quad + \sum_{r=1}^{n}\frac{\partial^2 f_i(\bar{x},\theta,w)}{\partial\bar{x}_r\partial\theta_j}\frac{\partial\bar{x}_r}{\partial\theta_k} + \frac{\partial^2 f_i(\bar{x},\theta,w)}{\partial\theta_k\partial\theta_j},
\end{aligned}
\tag{3.50}
$$

leading to a linear equation for the second order sensitivity $\widetilde{S}_i^{\;jk} = \partial_{\theta_k} S_i^{\;j}$,

$$
\begin{aligned}
0 = \sum_{r=1}^{n}\sum_{l=1}^{n}&\left(\frac{\partial^2 f_i(\bar{x},\theta,w)}{\partial\bar{x}_r\partial\bar{x}_l}S_r^{\;k} + \frac{\partial J_f(\bar{x},\theta,w)_i^{\;l}}{\partial\theta_k}\right)S_f(\bar{x},\theta,w)_l^{\;j}\\
&+ \sum_{l=1}^{n} J_f(\bar{x},\theta,w)_i^{\;l}\widetilde{S}_f(\bar{x},\theta,w)_l^{\;jk} + \sum_{r=1}^{n}\frac{\partial^2 f_i(\bar{x},\theta,w)}{\partial\bar{x}_r\partial\theta_j}S_f(\bar{x},\theta,w)_r^{\;k}\\
&+ \frac{\partial^2 f_i(\bar{x},\theta,w)}{\partial\theta_k\partial\theta_j}.
\end{aligned}
\tag{3.51}
$$

Again, the existence of a solution depends on the invertibility of the Jacobian $J_f(\bar{x},\theta,w)$. We note that the same LU-decomposition of J_f can be used for the numerical solution of (3.49) and (3.51). Usually all derivatives of f appearing in (3.48) and (3.51) can be calculated analytically; for large systems, a symbolic calculation package will do the job, e.g. GiNaC, GNU OCTAVE-forge's symbolic package or MATLAB's symbolic toolbox.

The solution of these linear systems of algebraic equations is usually a lot faster than the numerical integration of the ODE with sensitivity analysis.

4. Discussion on Tuning

None of the MCMC methods discussed here are free of internal parameters, which tremendously influence the sampling efficiency. We will call them *tuning parameters*. These parameters often set the average distance between successive points in a Markov chain and, in turn, the acceptance rate; these properties change the convergence speed. Badly tuned algorithms force the user to wait longer for the same amount of information about the target probability density since larger samples need to be collected. The tuning of an algorithm is a difficult optimisation problem in itself, but fortunately need not be solved exactly for the sampling to be a valid approach.

When comparing two competing algorithms, it is important to tune both of them for high efficiency; otherwise it would be easy to produce arbitrary results.

4.1. Comparing Two Algorithms from Different Families Requires Careful Tuning

The smaller the number of tuning parameters, the easier it is to adjust the algorithm. Most algorithms used throughout this manuscript have exactly one scalar tuning parameter $b \in \mathbb{R}_+$, so the tuning procedures can be fairly similar. One exception are Hamiltonian type algorithms, which require a number of steps within the transition trajectory, so the size (real) and number (integer) are both algorithm parameters in this case. Another example for an integer tuning parameter can be found in DRM and DRAM: both use a positive integer for the number of rejection delays. For these parameters sensible values are not difficult to find and we do not subject them to optimisation.

It is very useful to know some rules for the scaling of these parameters with problem size. Such rules can be found in literature for many well known algorithms (Beskos et al. 2013; Roberts and Rosenthal 2001). The comparison of several vastly different algorithms in development is more difficult.

The average acceptance rate within the Markov chain is a sensible diagnostic, it is calculated using the sample $\Theta(b;\theta_0)$ (or a subsample of fixed length), but depends on the tuning parameter b which was used to obtain the sample. We abbreviate the symbol $a(\Theta(b;\theta_0))$ to simply $a(b)$ for convenience. For well understood algorithms a sensible value for $a(b)$ is known from the literature. Setting b appropriately, while monitoring $a(b)$ results in good performance. A caveat of this approach is that the average acceptance will change depending on the region in parameter space and progression of *convergence*. Good acceptance *initially* (early on during sampling) means fast convergence, but for a representative sample acceptance needs to be good when the Markov chain reaches one of the modes (maxima) of the target density, the location of which is not known beforehand. Bad initial acceptance does not mean bad performance when converged.

For these reasons a different strategy during convergence is advisable. Since, in fact, average acceptance will change slightly depending on the local environment of the current parameter vector, it is not sensible to calculate gradients $\nabla_b a(b)$, making the optimisation more difficult. The tuning procedure is usually iterative and is performed using consecutive, adjoining sub-samples, which will be referred to as windows.

The acceptance rate is averaged over a sufficiently large window of successive steps and the appropriate window size also depends on the auto-correlation length. This window size s may become an additional unknown parameter that needs to be adjusted. As an example, consider auto-correlation lengths $\tau_{\text{int.},\ell}$ of order 10^3 and an acceptance averaging window of $s = 10^2$, the estimate

$$[a(b);s] := 1 - \frac{1}{s}\sum_{j=i}^{i+s} \delta_{\theta_{j+1},\theta_j}\,, \qquad \theta_j, \theta_{j+1} \in \Theta_b \tag{4.1}$$

will become problematic due to its small size. Since $\tau_{\text{int.},\ell}$ is very large, the Markov chain hardly moves during this averaging process. So, the average is not trustworthy, it is merely a snapshot and will change if i is shifted by more than $2\lceil \tau_{\text{int.},\ell} \rceil$. Clearly, we have to increase s and possibly reduce it again (for efficiency) once b is closer to optimal (small $\tau_{\text{int.},\ell}$). Of course, this approach becomes problematic if the user simply does not know appropriate acceptance values for the chosen algorithm.

Some algorithms are tuned by choosing a particular transition probability density, thus making this probability's parameters the ones to tune. Whenever the choice is free, we will use multivariate Gaussians $\mathcal{N}(\mu, \Xi)$; in logarithmic space, it seems a sensible default. In some cases, where the structure of Ξ is not specified by the algorithm, we chose a diagonal or even unit matrix structure, i.e. $I_m b$. In all cases discussed here, μ is specified entirely by the algorithms we use, for Metropolis type algorithms $\mu = \theta_i$.

Table 4.1 summarises the most important tuning parameters and their interpretation within the respective algorithm.

To get a measure of sampling quality, we define a quantity which takes many different effects into account, the *effective sampling speed*:

$$v_{\text{eff.},\ell} = \frac{N}{2\tau_{\text{int.},\ell} t_{\text{s}}}, \tag{4.2}$$

where t_{s} is the CPU-time, i.e. duration of the sampling process (ℓ refers to the observable for which auto-correlation is computed, in our case it is the log-posterior). We will use this definition whenever we obtain a large sample after a sufficiently long *burn-in* phase.

The planning of the burn-in procedure can be done in various ways, often we can only decide whether the burn in was sufficiently long after the fact (Brooks and Gelman 1998; Cowles and B. P. Carlin 1996; Gelman and Rubin 1992; Plummer et al. 2006).

For all samples used in this manuscript we have compared the length of the burn-in phase N_0 to the posterior sample's auto-correlation length $\tau_{\text{int.},\ell}$. If $N_0 \gg \tau_{\text{int.},\ell}$, then the *post burn in position* of the Markov chain is independent of the initial position, so the chain is likely converged. We also inspect the log-posterior values visually and if there is a noticeable drift, we increase N_0 and sample size N.

It is often better to avoid the measurement of run-times when they are expected to be very short, because of overhead in function calls. Depending on the implementation and cpu architecture there might be a big difference between wall clock time and cpu time, the latter being the correct one to estimate computation costs. For this reason we have devised an optimisation cost function that does not require time measurements, but still works for generic algorithms, because it can incorporate the users knowledge about the algorithm. The resulting tuning method is described in Section 4.2; for the results in this manuscript we have used both this method and the technique of acceptance tuning.

algorithm	tuning par.	type	meaning (short)
MH	ϵ	$\mathbb{R}_{++}$	transition covariance scale
DRAM	nd	$\mathbb{N}$	number of delays
	$s(m)$	$\mathbb{R}_{++}$	covariance factor
	ϵ	$\mathbb{R}_{++}$	invertibility term
SMMALA	ϵ	$\mathbb{R}_{++}$	move distance
HMC	b	$\mathbb{R}_{++}$	integration step size
	n_{steps}	$\mathbb{N}$	number of steps during transition
RMHMC	b	$\mathbb{R}_{++}$	integration step size
	n_{steps}	$\mathbb{N}$	number of steps during transition
	n_{FP}	$\mathbb{N}$	number of fixed point iterations

Table 4.1.: Meaning of the tuning parameters for the discussed algorithms

4.2. Algorithm-Independent Tuning

As discussed in the previous section, it is important to tune competing algorithms equally well. Otherwise it is easy to skew the results in favour of one algorithm or another by simply misadjusting one set of parameters.

Here, we suggest a generic tuning algorithm that works equally well for all sampling algorithms, but focuses on a particular observable (a variable that depends on the sample members θ_i, e.g. $\ell(\theta;\mathcal{D},M)$). This pragmatic and robust optimisation scheme also returns some measure of tuning uncertainty when done. The quantity we chose to optimise is not the acceptance rate but a mixture of a and the auto-correlation length $\tau_{\text{int.},\ell}$. We want auto-correlations to be small, because large values decrease the effective sample size and therefore the effective speed of any algorithm. The effective sample size

$$N_{\text{eff.}} = \frac{N}{2\tau_{\text{int.},\ell}} \tag{4.3}$$

can be understood as the equivalent size of an uncorrelated sample (Wolff 2004) which provides the same quality of estimation (standard deviation) of the observable in question (ℓ).

Yet, for some algorithms auto-correlation free samples come at the high price of increased computation time, this is the case for Hamiltonian type algorithms, for others $\tau_{\text{int.},\ell}$ cannot be reduced below a minimum value. Clearly, just minimising $\tau_{\text{int.},\ell}$ won't suffice if computation time is considered.

But, as previously mentioned, time measurements have some drawbacks as well. We can extend this point further:

- When model parameters are chosen badly, the model simulation time might fluctuate a lot between two bad parametrisations as they might lead to very different model behaviours, e.g. stiffness.
- An ODE integrator might give up on certain parameters and return quickly, seemingly decreasing the sampling time.
- Time measurement functions differ depending on the chosen software. the functions `tic` and `toc` in matlab will return different values than the function `cputime` does.
- In `C` many different time measurement libraries exist.

But, if comparability between implementations is not required, $v_{\text{eff.},\ell}$ can still be used as an optimisation cost function.

In our generic tuning procedure, we make the assumption that the user knows about the computational costs or other drawbacks of increasing a tuning parameter. This knowledge can be defined using a prior probability density for the tuning parameter. If large values are not desired (e.g. a large number of HMC steps), the user can assign very low probabilities to them.

The average acceptance rate on the other hand is valuable because of the difficulty to estimate large $\tau_{\text{int.},\ell}$ within short chains. For close to zero acceptance ($0 \leq a \ll 1$) $\tau_{\text{int.},\ell}$ becomes very large, at $a = 0$, $\tau_{\text{int.},\ell}$ is impossible to estimate. This becomes evident in huge standard deviations for $\tau_{\text{int.},\ell}$. Needless to say, it is very impractical to calculate numerical derivatives of $\tau_{\text{int.},\ell}$ and a via finite differences, when the values have significant noise.

These considerations lead us to construct a sampling technique for both optimisation and convergence purposes, which can be described as *Bayesian tuning*. We assign a prior distribution $p(b)$ to the tuning parameter based on our experience and knowledgein terms of standard distributions: Gamma, Beta, Gaussian, Poisson, etc. Then, we sample from the tuning-prior directly using the appropriate random number generator to obtain a list of candidate b values: $\{b_j; j = 1, \ldots, n_b\}$. We then obtain a quality rating or *weight* $w(b_j)$

for each b in the list. This is done by simulating a test Markov chain:

$$b_j \mapsto \tau_{\text{int.},\ell}(b_j), a(b_j), \tag{4.4}$$

$$w(b_j; \tau_{\text{int.},\ell}, a) = \gamma \exp\left(-\gamma\, \tau_{\text{int.},\ell}(b_j)\right) a(b_j), \tag{4.5}$$

$$p(b_j|\tau_{\text{int.},\ell}, a) \propto w(b_j; \tau_{\text{int.},\ell}, a) p(b_j), \tag{4.6}$$

where $a(b_j)$ is the average acceptance rate for the chosen b_j. Our choice for the weight function w in (4.5) is motivated by the following considerations, inspired by the convergence properties of Markov chains (Madras and Randall 2002; Wolff 2004):

- Convergence to the target distribution is exponential, with scale τ.
- The typical MCMC run-time rate of convergence τ is approximately proportional to $\tau_{\text{int.},\ell}$.
- The correlations between members of the Markov chain decrease exponentially with the number of steps.
- Very low acceptance values make $\tau_{\text{int.},\ell}$ estimation very inaccurate.

The estimator of $\tau_{\text{int.},\ell}$ in Wolff (2004) returns values that suggest no auto-correlation at all for inputs with an acceptance of 0%. In that case the factor of $a(b_j)$ in (4.6) will correctly set the weight $w(b_j)$ to zero. Large auto-correlations and low acceptance rate will both result in very small weights. This makes the weighted average robust with respect to the estimation of $\tau_{\text{int.},\ell}$. Definition (4.5) in this context has similarities to a likelihood function and fits well into the language of Bayesian analysis. This is why we have elected to use the posterior density $p(b|\tau_{\text{int.},\ell}, a)$ as the quality indicator for tuning. The free parameter γ can be chosen by the user, it should reflect the scale of $\tau_{\text{int.},\ell}$ (if $\log_{10}(\tau_{\text{int.},\ell}) \approx s$ then γ should be close to 10^{-s}). Its practical role is to also mediate between the two considered characteristics of a Markov chain, acceptance and auto-correlation.

We chose to do R simulations in parallel and then select the endpoint of the chain with the highest $w(b_j)$ value as a starting point for R new chains. In this way, we also progress towards convergence during the tuning process. This can be iterated as many times as necessary. On each iteration, our implementation updates graphs for $a(b)$, $\tau_{\text{int.},\ell}(b_j)$ and the current best value for b: the weighted average

$$\hat{b} = \frac{\sum_j w(b_j; \tau_{\text{int.},\ell}, a) b_j}{\sum_j w(b_j; \tau_{\text{int.},\ell}, a)} = [b] . \tag{4.7}$$

This estimator is equivalent to calculating the posterior mean from (4.6). We choose to calculate the mean instead of the mode as this is more convenient with regard to calculations of error propagation; i.e.: to calculate regions of confidence for $\hat{b}$. The advantage of this approach is that it is very close to the thought process of the user of MCMC algorithms as Bayesian approaches often are. Most tuning procedures require the user to guess regions with suitable parameters. So, we expect the user to have reasonable prior information on b, which is then updated through observation, as it should be.

This generic method works equally well for all mentioned algorithms. The process is illustrated for the Metropolis algorithm in Figure 4.1. In the top frame, we have included $\hat{b}$; it is shown as a filled square accompanied by two filled triangles indicating the estimation uncertainty. The acceptance rate for the metropolis algorithm is here about 23% and agrees very well with values given in literature (e.g.: Gelman, Roberts, and Gilks 1996).

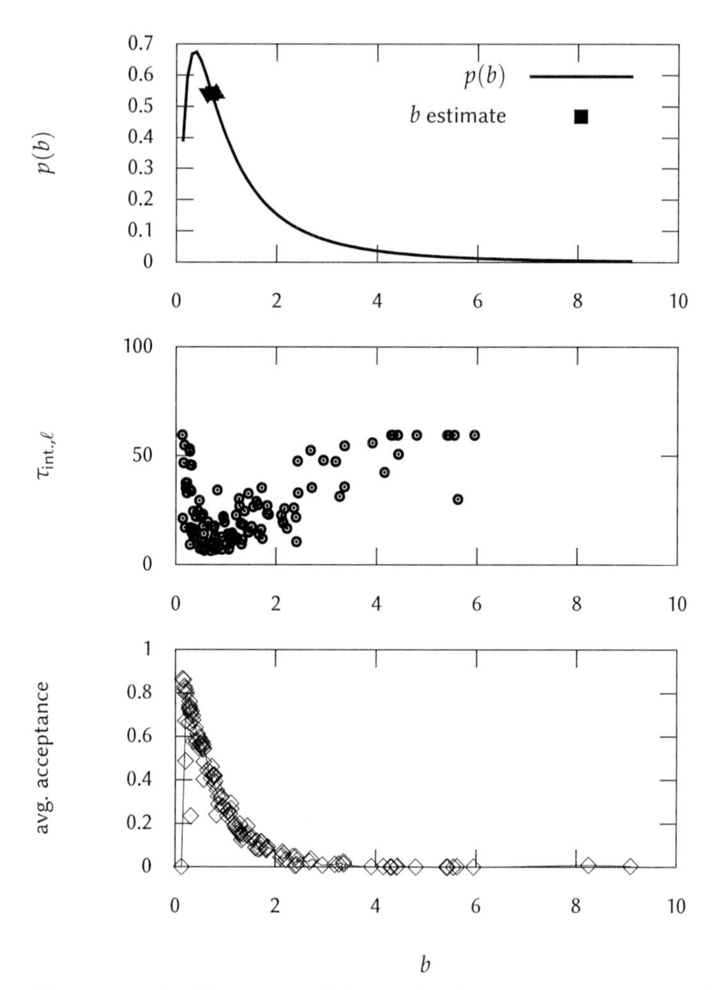

Figure 4.1.: An illustration of the results that generic tuning provides. Here, we tuned the step size for the standard Metropolis algorithm. Shown are the prior (top, line) for the tuning parameter b as well as (mid) the autocorrelation length used as sample quality diagnostic (4.5) and the average Markov chain proposal acceptance (bottom).

5. Algorithm Development

The convenient properties of steady states open new possibilities to speed up sampling algorithms. Starting from SMMALA and RMHMC, we can replace the numerically expensive sensitivity analysis and the integration of ODES by much cheaper calculations, which only work in that case. The following Section introduces two different methods to calculate steady states, which we have used within the RMHMC and SMMALA algorithms, while Section 5.2 deals with a method of efficient estimation techniques for the sensitivities and metric tensor $G(\theta)$. We will refer to algorithms which make use of the metric tensor $G(\theta)$ as *advanced* in contrast to algorithms such as the (adaptive) Metropolis-Hastings algorithm, because the former use more of the ODE model's mathematical structure to adapt while the latter are more generic.

Each section will introduce a numerical technique as well as a description of the MCMC algorithm derived from that technique. As an example, we will use the SMMALA algorithm variant each time but the changes can be applied to any algorithm, e.g. RMHMC and the Metropolis-Hastings algorithm family. A more in-depth discussion of the RMHMC variants can be found in Kramer, Calderhead, and Radde (2014).

In Section 5.3 we present a comparison of different algorithm subsets listed here.

5.1. Calculation of Steady States

The availability of steady state sensitivities makes it easy to estimate the shape of the bifurcation branch of the steady state. This means that we can update the steady states when small parameter changes occur. This is of great value during trajectory integration of Hamiltonian-type algorithms and to a lesser degree for MALA and Metropolis type algorithms. Whenever a small parameter change occurs, the first order Taylor approximation of the new steady state is:

$$\bar{x}(\theta + \Delta_\theta, u_k) \approx \bar{x}(\theta, u_k) + S_f(\bar{x}, \theta, u_k)\Delta_\theta \,. \tag{5.1}$$

This approximation breaks down in the vicinity of bifurcation points, where additional effort is required, depending on the type of bifurcation.

Having a good estimate available, it would be wasteful to recalculate the steady states from the initial conditions using the ODEs flow. Instead we have initialised the steady state variable using a numerical ODE solver and thereafter used the Newton-Raphson method to correct the initial guess (5.1).

5.1.1. Newton-Raphson Method

Given a differentiable function f with a vector argument x (with an initial guess), we can approximate $\bar{x}$ iteratively such that $f(\bar{x}) = 0$ using the Newton-Raphson method:

$$x_{i+1} = x_i - J_f(x_i)^{-1} f(x_i)\,, \tag{5.2}$$

where J_f is the Jacobian of f (i is the iteration index, with x_0 being the initial guess). This is precisely the problem we have to solve repeatedly whenever experimental data is known to be in steady state. The code we used for this purpose is printed in Listing 5.1.

Listing 5.1: GNU Octave/Matlab code of the Newton-Raphson method

```
 function [xs]=newton_raphson(F,X,rho,U)
%
% [xs]=newton_raphson(F,X,rho,U)
%
% finds the steady states of ode system in F
% one steady state per entry in U{:}
% X{:} provides starting points, one per input
%
% F is a struct containing:
% .f: rhs of ode [function-handle]
%       f(x,rho,u)
% .Jf: Jacobian of f [function-handle]
% .Sf: Sensitivity of steady state [function-handle]
%        dx(rho,u)/drho
%
% returns xs{:}
%        one steady state xs{i} per input U{i}
%

NumOfObs=length(U);
% tolerances for x and f
rtol=1e-6;
```

```
atol=1e-6;

iterations=0;
maxIterations=40;

for i=1:NumOfObs
    x=X{i};
    dx=x;
    while (any(abs(dx) > rtol*abs(x)+atol) &&
                              iterations < maxIterations)
     dx=-F.Jf(x,rho,U{i})\F.f(x,rho,U{i});
     x=x+dx;
     iterations=iterations+1;
    end%while

    if any(abs(F.f(x,rho,U{i})) > rtol*abs(x) + atol)
        xs{i}=x;
        warning('[...]');
    else
        xs{i}=x;
    end%if
end%for

end%function
```

5.1.2. NR+SMMALA for Models with Steady State Observations

This algorithm was published in Kramer, Calderhead, and Radde (2014) and utilises the Newton-Raphson method to calculate the model response, but huge numerical efficiency is drawn from efficient sensitivity calculations as described in Section 3.8.

1. initialise Markov chain using an ODE solver or an analytical steady state solution if one is available for a specific set of plausible parameters $\theta_0 \rightarrow \bar{x}(\theta_0, u_k)$
2. initialise metric tensor $G(\theta_0)$
3. initialise sample counter $i := 0$ and sample container $\Theta := \varnothing$
4. repeat the following steps until sample counter reaches final value
   a) solve linear system for S_f, build metric tensor $G(\theta_i)$ and its Cholesky factor

b) calculate mean $\mu(\theta_i, \epsilon)$ of SMMALA transition density $p(\phi|\theta_i)$ according to (3.39)

c) calculate covariance $C^{\mathrm{sm}}(\theta_i, \epsilon)$ of $p(\phi|\theta_i)$ (3.40)

d) draw a proposal ϕ from $\mathcal{N}(\mu(\theta_i, \epsilon), C^{\mathrm{sm}}(\theta_i, \epsilon))$

e) shift current steady states $\bar{x}(\theta_i, u_k)$ using steady state sensitivities (5.1):

$$\xi_k := \bar{x}(\theta_i, u_k) + S_f(\bar{x}, \theta_i, u_k)(\phi - \theta_i) \tag{5.3}$$

f) use ξ_k as initial condition for the Newton-Raphson method to obtain $\bar{x}(\phi, u_k)$

g) obtain model response $h(\bar{x}(\phi, u_k), \phi, u_k)$

h) repeat step 4a to 4c using proposal ϕ as given (instead of θ) for the reversal properties $\mu(\phi, \epsilon)$ and $C^{\mathrm{sm}}(\phi, \epsilon)$ of this transition

i) calculate acceptance probability $\alpha(\phi|\theta_i)$

j) draw a uniform random number $r \in (0, 1)$, append proposal ϕ to the sample Θ if $r < \alpha(\phi|\theta_i)$ (i.e. *accept*) otherwise append θ_i to the sample Θ again (i.e. *reject*)

k) increase sample counter

5. return Θ and each sample member's posterior probability value.

5.1.3. Circuit Breaking Algorithm

The circuit breaking algorithm (CBA) is described in Radde (2010) and is applicable to a broad range of ODE models but originally formulated for input free systems. This method is best explained within the directed graph representation of ODE models, as described in Section 2.2.3. The first step of the CBA is to transform the models graph into a DAG. We do this by identifying a minimal set of vertices which are members of all cycles and cut all influx to them: remove all inward pointing edges. This breaks the cycles. Subsequently, these nodes are assigned adjustable values, parameters κ_j. In a sense, these vertices are replaced by the κ_j.

Parametrised intermediate solutions for the steady states can be found by solving implicit equations of the form $f_i(\bar{x}(\kappa)) = 0$ in topological order. Here, some of the $\bar{x}_i(\kappa)$ are still replaced by the κ_j parameters, for example:

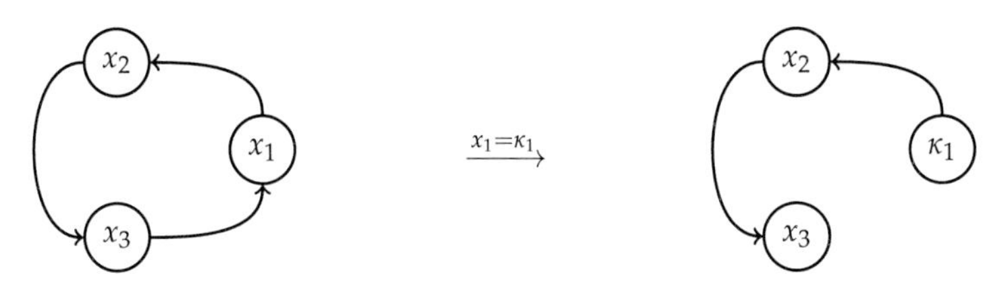

In this example, the solutions can be found by first solving $f_2(\bar{x}_1 = \kappa_1, \bar{x}_2) = 0$ for $\bar{x}_2$ and then $f_3(\bar{x}_2(\kappa_1), \bar{x}_3) = 0$ for x_3. Every solution $\bar{x}(\kappa)$ has to be finalised by finding the κ-zeros of the circuit characteristic $c(\kappa) = f_1(\kappa_1, \bar{x}_2(\kappa_1), \bar{x}_3(\kappa_1))$. This action can be viewed as a restoration of the original graph topology. In the special case where removing one vertex breaks all cycles in the model's graph we can derive a very efficient algorithm for steady state calculations and apply it during the sampling. The more general case of more than one removed vertex is discussed in greater detail in Radde (2010).

In systems biology, ODE models are most commonly constructed from rational functions. Therefore, the circuit characteristic may very well be a rational function as well. However, there are three major obstacles to this approach:

1. A very common experiment model is a step function (or sigmoidal function), it describes the on switch of the experiment. Step functions are not rational and can complicate the approach. In such cases input dynamics have to be set to their asymptotic $t \to \infty$ value.

2. Square roots are not rational, but can occur in this framework (fractional exponents in general).

3. When rapidly testing hundreds of different models, some might have the necessary conditions for a successful calculation of $c(\kappa)$ while others might not.

A minor point is that the necessary analytical computations often require familiarity with a symbolic computation software, while the sampling itself requires familiarity with numeric calculation software.

But, in the fortunate case of a rational circuit characteristic, finding numerical approximations for its zeros is extraordinarily easy. In one dimension (scalar $\kappa = \kappa_1$), it is possible to transform the zero problem into an eigenvalue problem via the companion matrix representation. The zeros (roots) of the polynomial

$$g_a(x) = x^4 + a_3 x^3 + a_2 x^2 + a_1 x + a_0 \tag{5.4}$$

can be computed as follows:

$$A = \begin{pmatrix} -a_3 & -a_2 & -a_1 & -a_0 \\ 1 & 0 & 0 & 0 \\ 0 & 1 & 0 & 0 \\ 0 & 0 & 1 & 0 \end{pmatrix}, \qquad \lambda_a = \mathrm{eig}(A) \tag{5.5}$$

$$0 = \det(A - I_4 \lambda_a) \stackrel{!}{=} g_a(\lambda_a) \tag{5.6}$$

where the exclamation mark signifies an identity due to the purposeful construction of the companion matrix A. This technique allows rapid calculation of steady states at almost negligible cost.

Let's consider the CBA on the following example:

$$\begin{aligned} \dot{x}_1 &= \rho_1 u_1 + \frac{1}{10} x_1 x_2 - x_1\,, \\ \dot{x}_2 &= \rho_2 + \frac{1}{5} \frac{x_2 x_3}{1 + x_3} - x_2\,, \\ \dot{x}_3 &= \rho_3 + \frac{1}{1 + x_1^2} - x_3\,. \end{aligned} \tag{5.7}$$

This model has one cycle, containing all vertices. We can break it by replacing: $x_1 \stackrel{!}{=} \kappa_1 = \kappa$. This leads to the following κ dependent steady state condition

$$\begin{aligned} \bar{x}_1(\kappa) &= \kappa\,, \\ \bar{x}_2(\kappa) &= -\frac{5\rho_2(\bar{x}_3(\kappa) - 4)}{\bar{x}_3(\kappa)}\,, \\ \bar{x}_3(\kappa) &= \rho_3 + \frac{1}{1 + \kappa^2}\,. \end{aligned} \tag{5.8}$$

The circuit characteristic $c(\kappa) = f_1(\bar{x}(\kappa)) = 0$ is given by

$$c(\kappa) = \rho_1 u_1 - \frac{1}{2}\kappa \left(\rho_2 - 4 \frac{\rho_2(1 + \kappa^2)}{\rho_3(1 + \kappa^2) + 1} \right) - \kappa\,. \tag{5.9}$$

This characteristic can be decomposed into a nominator polynomial and a nonzero denominator polynomial in κ. Given any particular value for ρ, (5.5) enables us to immediately calculate the zeros of $c(\kappa)$ and thus the steady state $\bar{x}$, for any given model parameter vector ρ. Figure 5.1 shows a bifurcation diagram, that was calculated using the CBA, with ρ_2 used as the bifurcation parameter. We have only used the circuit characteristic (an algebraic function). No model integration was performed, rendering this method faster and simpler than standard bifurcation analysis. For some parameters the model might converge only very slowly to the asymptotic steady state, which requires longer integration times, but the CBA will compute the result quickly independent of parametrisation and model sluggishness. We can use this to our advantage whenever steady state data is used in Bayesian parameter estimation. This method can be used in conjunction with any of the previously described algorithms. To demonstrate this, We have combined the SMMALA algorithm with this method and added an output function and experiment description to (5.7):

$$\begin{aligned} h(x) &= \begin{pmatrix} x_1 + x_2 \\ x_2 + x_3 \end{pmatrix}, \\ u &\in \{0, 0.4, 0.6, 0.8, 1.0, 1.2\}. \end{aligned} \tag{5.10}$$

We have used this output function to numerically simulate data using forward integration and normally distributed random numbers. The resulting sample Θ can be used to obtain output trajectories. Figure 5.2 shows the output function values $\Theta \mapsto \{h(\bar{x}(\rho_i, u), \rho_i, u) : i = 1, \dots, N, \rho_i = exp(\theta_i)\}$ for each sample member along with the data.

5.1.4. CBA+SMMALA for Models with Steady State Observations

This SMMALA based algorithm uses the CBA for steady state calculations and benefits from efficient sensitivity analysis.

1. Use symbolic calculations to break all cycles in the model's graph and compute the circuit characteristic
2. decompose the circuit characteristic in a nominator $g_\mathrm{N}(\kappa)$ and a denominator polynomial $g_\mathrm{D}(\kappa)$
    - *if* this fails abort this approach and use a more generic sampler instead (e.g. NR-SMMALA)

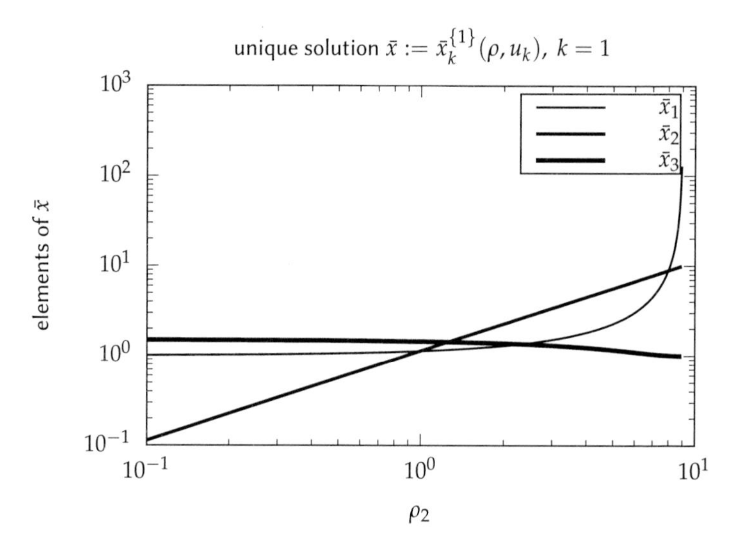

Figure 5.1.: Rapid steady state recalculations for example model (5.7) using the `roots` function of GNU OCTAVE. Shown are the individual vector elements of a unique solution using a constant input with parameters: $u_1 = 1, \rho_1 = 1, \rho_3 = 1$. This bifurcation diagram is exact to machine precision due to the very accurate eigenvalue solver.

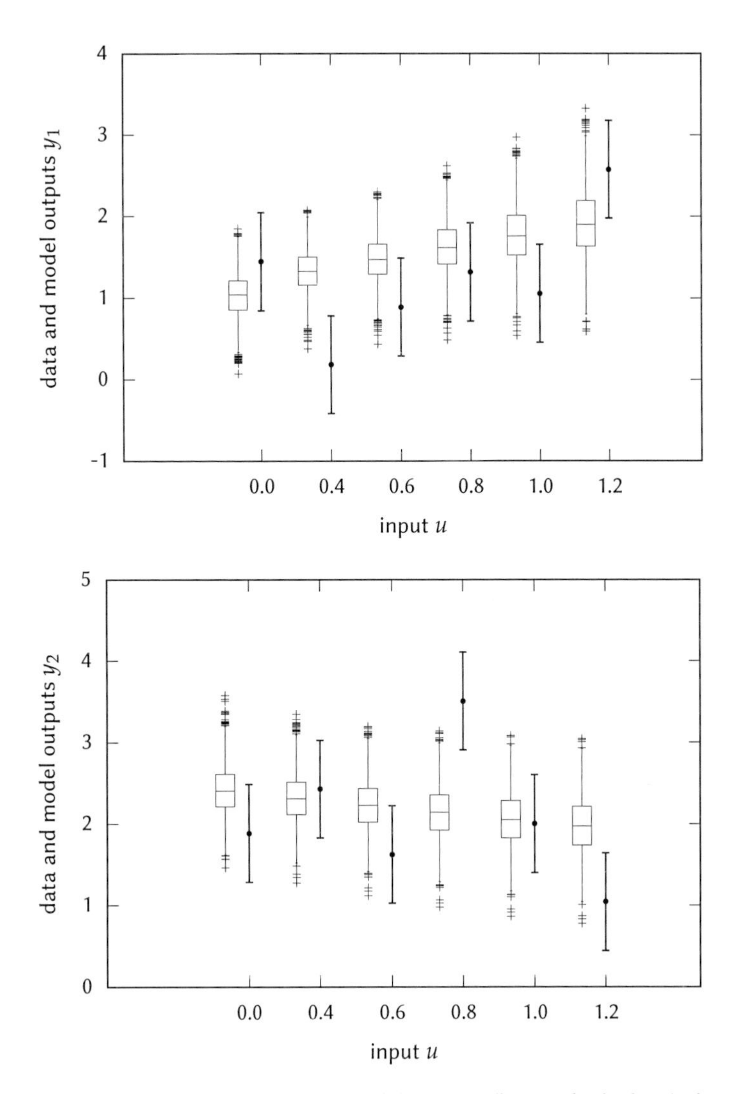

Figure 5.2.: Data (error bars) and model outputs (box and whiskers) obtained from a parameter sample (model (5.7)). Both outputs are shown, first (top) and second (bottom). The sample size was $N = 2^{16}$ and the effective sampling speed $v_{\text{eff.}} = 0.605(74)$ (average acceptance: 67%). The algorithm used SMMALA with circuit breaking steady state characterisation.

- *otherwise* construct a function `x.bar` which, given the coefficients of $g_{\mathrm{N}}(\kappa)$ and $g_{\mathrm{D}}(\kappa)$ will compute the zeros of $g_{\mathrm{N}}(\kappa)$ using the `roots` function: $\{\kappa^{\{l\}} : g_{\mathrm{N}}(\kappa^{\{l\}}) = 0\}$ and remove from this set all zeros of $g_{\mathrm{D}}(\kappa)$:

$$\{\kappa^{\{l\}} : g_{\mathrm{N}}(\kappa^{\{l\}}) = 0\} \setminus \{\kappa^{\{r\}} : g_{\mathrm{D}}(\kappa^{\{r\}}) = 0\}\,. \tag{5.11}$$

here «\» is the *set* operation `set-minus`

3. initialise Markov chain using an ODE solver or an analytical steady state solution if one is available for a specific set of plausible parameters $\theta_0 \to \bar{x}(\theta_0, u_k)$

4. initialise metric tensor $G(\theta_0)$

5. initialise sample counter $i := 0$ and sample container $\Theta := \varnothing$

6. repeat the following steps until sample counter reaches final value
   a) solve linear system for S_f, build metric tensor $G(\theta_i)$ and its Cholesky factor
   b) calculate mean $\mu(\theta_i, \epsilon)$ of SMMALA transition density $p(\phi|\theta_i)$ according to (3.39)
   c) calculate covariance $C^{\mathrm{sm}}(\theta_i, \epsilon)$ of $p(\phi|\theta_i)$ (3.40)
   d) draw a proposal ϕ from $\mathcal{N}(\mu(\theta_i, \epsilon), C^{\mathrm{sm}}(\theta_i, \epsilon))$
   e) obtain all steady states ξ_{jk} for θ_i from the `x.bar` function using the nominator and denominator of the circuit characteristic
   f) check all returned steady states for stability (e.g. via J_f), then *for each* stable steady state:
      i. calculate the output value $h(\xi_{lk}, u_k)$ for each stable steady state ξ_l
      ii. check the difference to the data, select the steady state that fits the data best: $\hat{\xi}_k = \arg\min_{\xi_{lk}} \|y_k - h(\xi_l, u_k)\|$
   g) obtain model response $h(\hat{\xi}_k, \phi, u_k)$
   h) repeat step 6a to 6c using proposal ϕ as given (instead of θ) for the reversal properties of this transition
   i) calculate acceptance probability $\alpha(\phi|\theta_i)$

j) draw a uniform random number $r \in (0,1)$, append proposal ϕ to the sample Θ if $r < \alpha(\phi|\theta_i)$ (i.e. *accept*) otherwise append θ_i to the sample Θ again (i.e. *reject*)

k) increase sample counter

7. return Θ and each sample member's posterior probability value.

5.2. Near Steady State Sensitivity Approximation

So far, all described techniques work well with steady state data, but not at all with time series data. We would like to provide an algorithm which bridges the gap between these two approaches.

The metric tensor does not need to be calculated precisely. An approximation that is cheaper to compute might result in better overall performance than the optimal, Fisher information derived metric. In this Section we will present such an approximation for time series data. Of course, this approximation gets worse when our assumptions diverge from the true model behaviour.

It is possible to obtain the sensitivities numerically at moderate costs, by using an ODE solver which performs sensitivity analysis. This is the default approach, which we employ in our C implementation of SMMALA, Section 6.1. But, given a fixed working environment (a set of programs, e.g. MATLAB), it can be hard to switch ODE solvers as a mere user. For example, OCTAVE's `lsode` function does not perform a sensitivity analysis during integration. To use it anyway, we can append equations to the state space:

$$\dot{x} = f(x,\rho,u)\,, \tag{5.12}$$

$$\dot{S}_f = J_f(x,\rho,u)S_f(x,\rho,u) + K_f(x,\rho,u)\,, \tag{5.13}$$

which can be solved for each column of S_f and K_f independently. This workaround can work well for small models and was used extensively in Girolami and Calderhead (2011) and Kramer, Calderhead, and Radde (2014).

Alternatively, we could try to approximate the values of $S_f(x,\rho,u)$ at each time point in a similar manner as for steady states. So, let us make some restrictions to the dynamics of the model, and calculate sensitivity approximations algebraically, at much lower cost than numerical integration.

We first solve (5.13) for constant $A = J_f$ and $B = K_f$:

$$\dot{S} = AS + B\,, \qquad S \in \mathbb{R}^{n \times m} \tag{5.14}$$

$$= A(S + A^{-1}B)\,, \qquad R = S + A^{-1}B \tag{5.15}$$

$$\Rightarrow \dot{R} = AR\,, \tag{5.16}$$

which is a linear system, with a known solution:

$$R(t) = \exp(At)R_0\,, \tag{5.17}$$

$$\Rightarrow S(t) + A^{-1}B = \exp(At)(S_0 + A^{-1}B)\,, \tag{5.18}$$

$$S(t) = \exp(At)(S_0 + A^{-1}B) - A^{-1}B\,. \tag{5.19}$$

At this point, using $S(t)$ as a blue print, we are free to insert the exact Jacobian $J_f(x(t), \rho, u)$ as well as $K_f(x(t), \rho, u)$ into solution (5.19) to use S as an approximation of the sensitivity matrices. For large t, such that $x(t)$ approaches a stable steady state, the approximation converges to the exact solution (3.48). For small t, correct initial conditions S_0 will make the approximation very similar to the exact solution. For known, parameter independent $x(t_0)$ initial conditions x_0, the initial parameter sensitivity S_0 is zero (a zero-matrix):

$$(S_0)_i^j = 0\,, \qquad S(t) = (\exp(At) - I)\,A^{-1}B\,. \tag{5.20}$$

The resulting approximation for the metric G can be used in both RMHMC and SMMALA. The drawback of this approach is that this metric does not describe the curvature of the parameter space exactly and the benefit of using a non-Euclidean metric might be lost altogether. This approach loses its validity entirely in cases like sustained oscillations (stable limit cycles). In other words, the systems behaviour should be monotone and bounded for best performance. We call the SMMALA variant derived from this method SSMMALA, it is defined in the following Section.

Using SSMMALA, we have obtained a parameter sample Θ for the model characterised by (5.7) and (5.10); we have used this model earlier to demonstrate the CBA. Using more simulated data points than before, shown in Figure 5.3, we have measured the effective sampling speed and compared it to that of unmodified SMMALA. Table 5.1 shows the results of this comparison, yielding an effective speedup of $11(2)$.

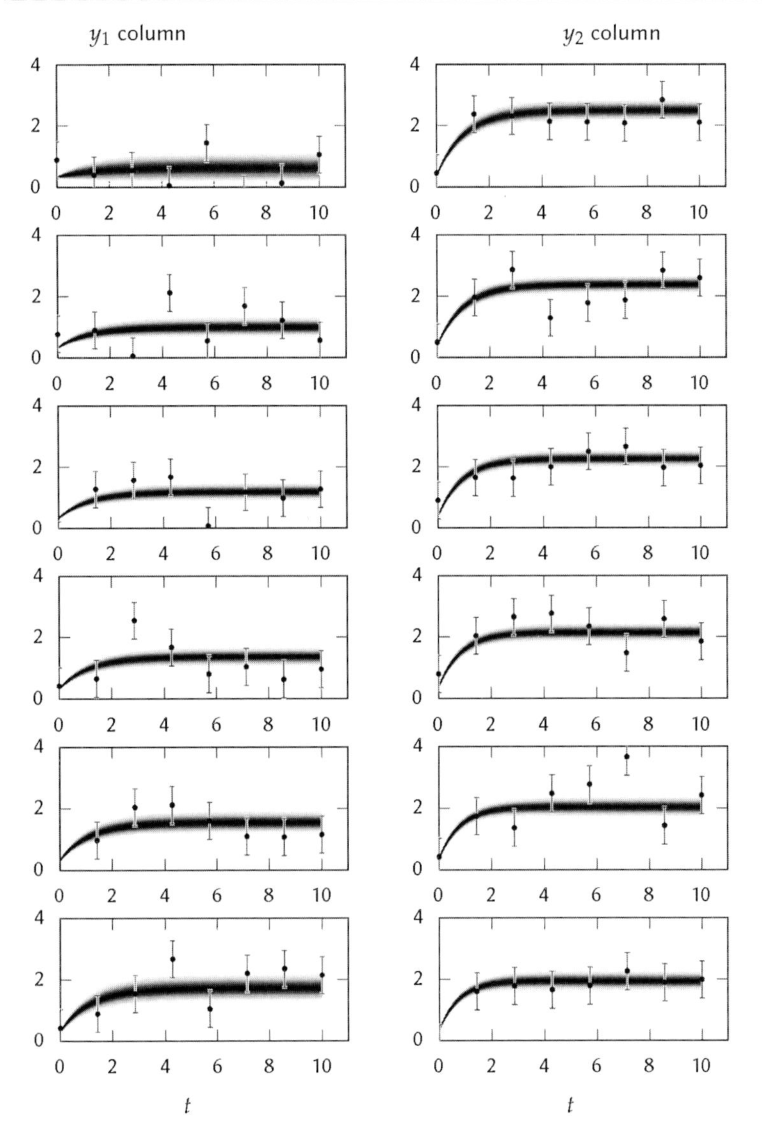

Figure 5.3.: Time series data fit using the sensitivity approximation approach for dynamic system (5.7) and output model (5.10). Shown are the posterior model trajectories colour coded by posterior density value and data as error bars. The data points were simulated using the model and Gaussian random numbers to simulate noise.

	SMMALA	SSMMALA
CPU cores used	8	3
sample size N	2^{16}	2^{16}
effective speed $v_{\text{eff.},\ell}$	$0.0054(5)\,s^{-1}$	$0.063(6)\,s^{-1}$
simulation time t_s	$77.2\,h$	$5.16\,h$
auto-correlation $\tau_{\text{int.},\ell}$	$21(2)$	$28(3)$

Table 5.1.: Juxtaposition of standard SMMALA and SSMMALA for (5.7). The speedup factor is $11(2)$. The first row lists the number of cores used by the ode solver built into GNU OCTAVE (this was determined by observing the CPU usage with top).

5.2.1. SSMMALA for Models with Time Series Data

The motivation for this algorithm is described in the previous Section. Instead of S_f, which is numerically found through sensitivity analysis during model integration, we use (5.20) for the calculations of the metric. The accuracy of the approximation determines the resulting algorithm performance. Otherwise we proceed similarly to SMMALA.

1. initialise Markov chain using an ODE solver or an analytical steady state solution if one is available for a specific set of plausible parameters $\theta_0 \to \bar{x}(\theta_0, u_k)$
2. initialise metric tensor $G(\theta_0)$
3. initialise sample counter $i := 0$ and sample container $\Theta := \varnothing$
4. repeat the following steps until sample counter reaches final value
   a) build metric tensor $G(\theta_i)$ using

$$S(t,x,\rho,u_k) = \left(\exp\left(J_f(x,\rho,u_k)t\right) - I_m\right) J_f(x,\rho,u_k)^{-1} K_f(x,\rho,u_k)$$
$$R_{jk} = \left.\frac{\partial h}{\partial x}\right|_{t_j,u_k}$$
$$G(\theta) := \sum_k \sum_j R^{\mathrm{T}} S(t_j, x_{jk}, \rho, u_k)^{\mathrm{T}} S(t_j, x_{jk}, \rho, u_k) R\,,$$

   as described in the previous Section.

b) calculate mean $\mu(\theta_i, \epsilon)$ of SMMALA transition density $p(\phi|\theta_i)$ according to (3.39)

c) calculate covariance $C^{\mathrm{sm}}(\theta_i, \epsilon)$ of $p(\phi|\theta_i)$, see (3.40)

d) draw a proposal ϕ from $\mathcal{N}(\mu(\theta_i, \epsilon), C^{\mathrm{sm}}(\theta_i, \epsilon))$

e) use any ODE solver for stiff problems to obtain $x(t_j, \phi, u_k)$

f) obtain model response $h(x(t_j, \phi, u_k), \phi, u_k)$

g) repeat step 4a to 4c using proposal ϕ as given (instead of θ) for the reversal properties of this transition

h) calculate acceptance probability $\alpha(\phi|\theta_i)$

i) draw a uniform random number $r \in (0, 1)$, append proposal ϕ to the sample Θ if $r < \alpha(\phi|\theta_i)$ (i.e. *accept*) otherwise append θ_i to the sample Θ again (i.e. *reject*)

j) increase sample counter

5. return Θ and each sample member's posterior probability value.

5.3. Discussion

We have introduced three techniques of model handling in conjunction with MCMC sampling: the circuit breaking algorithm in Section 5.1.3, sensitivity approximations in 5.2, and the Newton-Raphson method in 5.1.1. The latter approach and its *synergy* with sensitivity analysis is explained in more detail in Kramer, Calderhead, and Radde (2014). In this Section, we discuss the impact of the first two modifications. We compare SMMALA coupled with cba root finding (CBA+SMMALA), the use of sensitivity approximation for the computation of the metric (SSMMALA) and default SMMALA. Section 5.3.2 will extend this analysis to algorithms that use the Newton-Raphson method. This will include Hamiltonian type MCMC methods.

5.3.1. Comparing Different Data Cases Fairly

We have decided to test the performance of these approaches using a medium sized example of a biologically motivated model with $n = 6$ state variables and $m = 9$ parameters, as given in Listing B.2. Some more information on the model can be found in Appendix B.2. It represents the simplest hypothesis on the action of PKD in the vesicular transport of proteins from the trans-Golgi network to the cell membrane (Weber, Hasenauer, et al. 2011; Weber, Hornjik, et al. 2015). The simplified version of the PKD model that we use here has a scalar input variable $u \in \mathbb{R}$, which we set to the values

$$u_k \in \{0.4, 0.6, 0.8, 1.0, 1.2, 2\}, \tag{5.21}$$

with $k = 1, \ldots, 6$.

We use identical simulated, noisy time series data for SMMALA and SSMMALA but as CBA+SMMALA accepts only steady state data, the total number of data points is reduced for this algorithm. We use exactly the same output function in all three cases

$$y = \begin{pmatrix} x_1 \\ x_3 \\ x_5 \end{pmatrix}. \tag{5.22}$$

The two cases of *steady state* and *time series measurements* agree only in the respect that the system that generated the data used the same *true parameters*, which we shall endeavour to find. We expect the posterior distributions to be similar, but not identical. Here, we are interested in the goodness of the respective fits compared to the effort of finding them. The measurement time points for the time series data are $t_j \in \{1.5, 3.0\}$ ($T = 2$). The solution of the initial value problem in algorithms such as SMMALA and SSMMALA is the most costly part of all computations. Once a solution is obtained through numerical integration, the number of time points T to *compare* the simulation results and data *at* does not matter as this comparison is very fast. Adding different experimental conditions, i.e. increasing n_E, *will* require additional simulations and increase the overall numerical costs[1]. So, since we use identical experimental conditions u_k for all tested algorithms (CBA+SMMALA, SSMMALA and SMMALA) we have a fair comparison with regard to data processing. The difference in performance can only result from the differences in implementation or, possibly, from the different posteriors. Therefore, we must investigate the posterior shapes once sampling is done. It turns out that they are virtually identical.

Furthermore, this comparison makes sense even if we disregard computational complexity and even effective sampling speed or the finer details of data processing. Simply consider the quality of the fit against the time needed to obtain the fit, disregarding the uncertainty information for the moment. Figure 5.4 shows the fitting parameters. The *true* parameters are indicated by a dashed line. We have used them to generate the data. Visual

[1] so we don't do that

	$k = n_y \times T \times n_E$	χ^2_k	χ^2_k-CDF
CBA+SMMALA	$3 \times 1 \times 6$	18.7	0.5867
SSMMALA	$3 \times 2 \times 6$	28.6	0.1951
SMMALA	$3 \times 2 \times 6$	29.8	0.2448

Table 5.2.: χ^2_k statistic for the best fits of three MCMC algorithms, where k is the number of degrees of freedom. The interpretation of the CDF value is the probability to randomly obtain a smaller quadratic difference between observed data and the best fit of the model than the given χ^2 statistic if the data was model generated. Better fits result in lower CDF values.

inspection suggests that all three methods have recovered the parameters equally well. In practice, true parameters are unavailable, so we examine the goodness of fit. We have selected a best fit from each sample based on the highest recorded log-posterior (ℓ) value. The χ^2 statistic of these fits are listed in Table 5.2. The CDF values reveal that CBA+SMMALA has obtained a slightly worse fit than the algorithms which used more data, which is very reasonable and reassuring. Nevertheless, even for CBA+SMMALA's results, the CDF value suggests that there is a 60% probability to obtain a difference between data and fit *as extreme* as this *or lower* based on the error model. This value does not outright reject the possibility that the data could have been the result of random Gaussian noise *on the proposed model*. So, the quality of this fit is not bad by any means.

The sample generated by CBA+SMMALA is not vastly different from the other two, time series method generated counterparts, despite referring to different data (see Figure 5.4). But, the crucial point is that it comes at tremendously lower cost, since it was generated 10 times faster than SSMMALA (which itself was better than default SMMALA). The overall posterior shape for time series data can also be inspected in Figure 5.5, which reveals a unimodal density with strong correlations between some of the parameter pairs (a hexagonal binning method was used). The correlation coefficients are summarised in Figure 5.6. As mentioned before, strong correlations make adaptive methods necessary, otherwise acceptance drops to very low values and samples are highly auto-correlated. Markov chains that don't adapt to the target's shape are slower due to frequent rejection of moves. Which adaptation strategy is effective depends on the precise shape of the target density.

Adaptive Metropolis usually works very well for unimodal distributions of vaguely Gaussian shape. The normality of a sample can be tested in various ways, e.g. via the skewness and kurtosis (D'Agostino, Belanger, and D'Agostino Jr. 1990); the performance of AM however is not so sensitive as to warrant a precise test, a visual examination will suffice. Unfortunately, this information is available after the fact. We can test for normality or observe other properties of the sample only after sampling is complete and cannot a priori select the best algorithm for the task. So, given the posterior shape, it turns out to be not pressingly necessary to employ advanced algorithms such as SMMALA or RMHMC here.

It is important to note that adaptive Metropolis relies on the history of the Markov chain. This interferes with strategies such as population MCMC, where multiple chains are used simultaneously and chains can exchange positions. In these cases, the chains must exchange their histories. We discuss some of the more fundamental *drawbacks* of adaptive Metropolis in Section 5.3.3.

In all instances, our implementation language of choice, GNU OCTAVE, used all available CPU-cores (8 on the used machine[2]).

In summary: if the steady state problem can indeed be transformed into an eigenvalue problem for CBA+SMMALA, the benefits are huge. We see a speedup of 200 compared to the unmodified SMMALA for the PKD model. In the more general case, where the data is a time series, but the system nevertheless converges to a stable steady state, our modifications (SSMMALA) are still very useful and easy to implement. The speedup from standard SMMALA is still quite notable at about one order of magnitude.

5.3.2. Hamiltonian Type Algorithms Benefit Greatly from Speedups in the Model's Response

The speedup for Hamiltonian type algorithms such as RMHMC is far more dramatic. Since the model output and parameter space metric have to be recalculated many times for each MCMC transition step $\theta_i \to \phi$, any improvement to these computations is rewarded multiple times. Using three models of different size and real data[3], we have compared different implementations of RMHMC. The two larger models are discussed in Brännmark et al. (2010) and included in that publication's supplement. These models

[2] Intel® Core™ i7-4700MQ CPU @ 2.40 GHz

[3] For our purposes, this means that the systematic error Δ is not 0 and the notion of *true parameters* looses its usefulness.

deal with insulin signalling and test different hypotheses for the feedback structure and mechanism in the complex pathways that are triggered by insulin receptors. The input and output structures can be inspected in Listing B.5 and Listing B.6. As an example of a small model we have used MAPK signalling, defined in Listing B.4, which was published in the supplement of Fritsche-Guenther et al. (2011). The input model was added by us. Although the steady state data is given in arbitrary units, for comparison purposes, we treated it as fully quantitative and used a linear output function (Kramer, Calderhead, and Radde 2014). Figure 2.1 in Chapter 2 depicts the correctly normalised data as well as a fit obtained while taking normalisation into account.

We considered only the steady state data from Brännmark et al. (2010). We used computationally cheap, but exact steady state sensitivity analysis to compute the metric tensor in all methods labelled with an NR prefix as described in Section 3.8.

Table 5.4 shows relative speeds for Hamiltonian type algorithms ordered by problem size. The steady state specific variant NR+RMHMC is up to three orders of magnitude faster than the standard implementation of RMHMC. Omitted values for $\tau_{\text{int.},\ell}$ indicate that we managed to obtain only a few hundred sample points after many hours. In that case we resigned ourselves to upper bounds for the effective sampling speed as the samples are too small to calculate the auto-correlation $\tau_{\text{int.},\ell}$. The tests were done in MATLAB on a desktop computer with 12 cores[4]. We repeated these tests for Langevin type algorithms; the results are listed in Table 5.5. As can be expected, for NR+SMMALA our modifications are less rewarding then with NR+RMHMC, but still substantial at one order of magnitude. This is in very good agreement with the results in Table 5.3 which also show a tenfold speedup of SSMMALA versus SMMALA (but for time series data).

5.3.3. Adaptive Metropolis Rationale Is Sometimes Not Appropriate

As described in Section 3.5.3, the idea behind adaptive Metropolis is to estimate the target density's covariance and use a Gaussian probability density with a similar covariance as the Markov chain's transition kernel,

[4]Intel® Core™ i7-3930K CPU @ 3.20 GHz

	CBA+SMMALA	SSMMALA	SMMALA
$v_{\text{eff.},\ell}$ in s^{-1}	$4.41(36) \times 10^{-2}$	$4.07(36) \times 10^{-3}$	$2.22(58) \times 10^{-4}$
$v^{\text{rel.}}_{\text{eff.},\ell}$	199(54)	18(5)	1
acceptance	57%	52%	54%
$\tau_{\text{int.},\ell}$	72(6)	87(8)	83(22)
sample size N	2^{18}	2^{18}	2^{14}
CPU time t_s	11.4 h	102 h	124 h
wall clock time	1.45 h	15 h	15.7 h
step size ϵ	0.20	0.1847	0.1847

Table 5.3.: PKD model (Listing B.2): Comparison of three sampling algorithms using (auto-correlation corrected) effective sampling speed $v_{\text{eff.},\ell}$. All three algorithms are implemented in the GNU OCTAVE scripting language and run in the octave software. Time measurements were done using the `cputime` function, which returns the accumulated simulation time for all involved processors as well as `tic/toc` for wall clock time. On our machine 8 CPU cores were used, the elapsed wall clock time is about 8 times shorter. To make comparison most straightforward, we have used the same ϵ parameter for both SSMMALA and SMMALA since the posteriors are identical. This value agrees very well with our generic tuning algorithm (see Section 4.2).

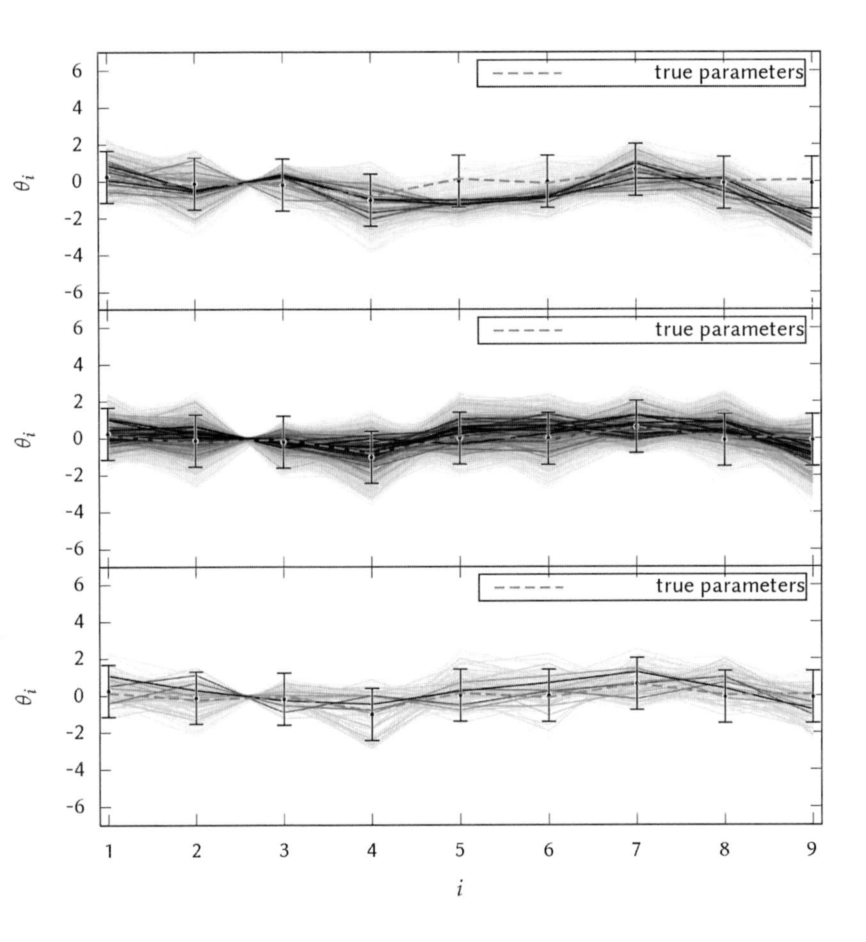

Figure 5.4.: Parameter samples for the PKD model obtained from the algorithms (from top to bottom): CBA+SMMALA, SSMMALA, SMMALA. For these plots, we skip $\lceil \tau_{\text{int.},\ell} \rceil$ points for every shown sample member. The number of lines is a visual cue for the effective sample size, which can be interpreted as the thoroughness of parameter space exploration. Darker lines correspond to higher posterior values.

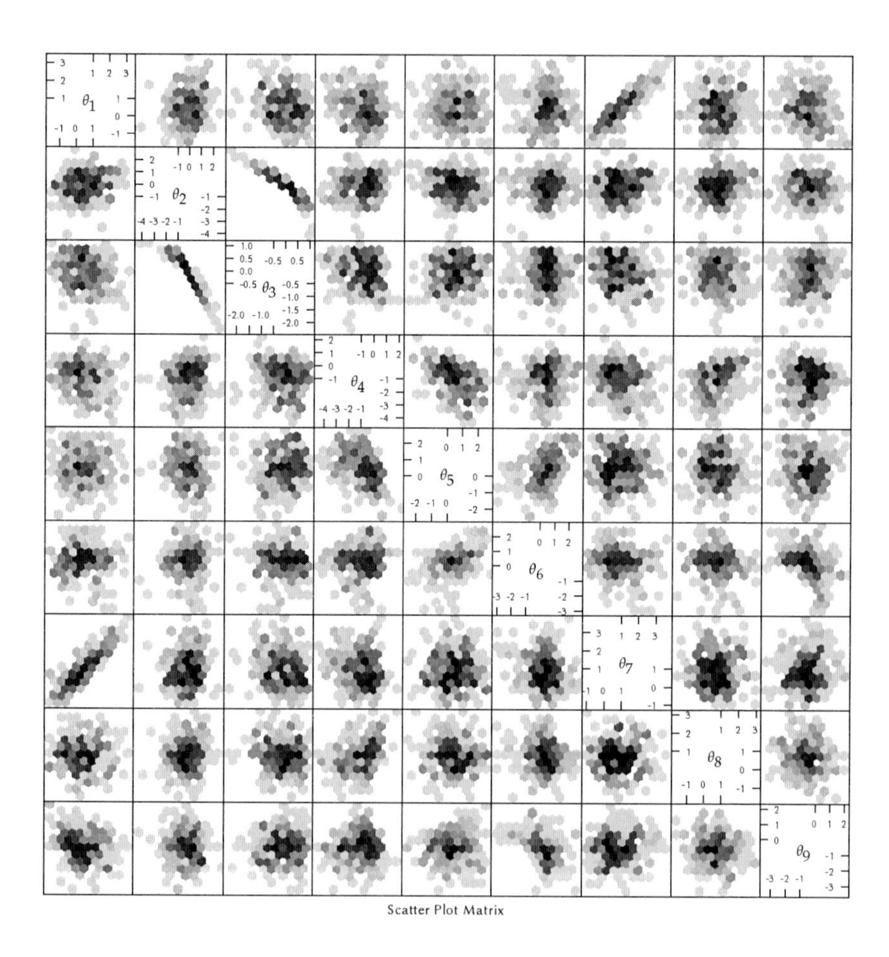

Figure 5.5.: A full correlation matrix plot of the posterior parameter sample for the PKD model in Listing B.2 obtained with SMMALA. We have used a hexagonal binning function to reduce clutter (R library: `hexbin`).

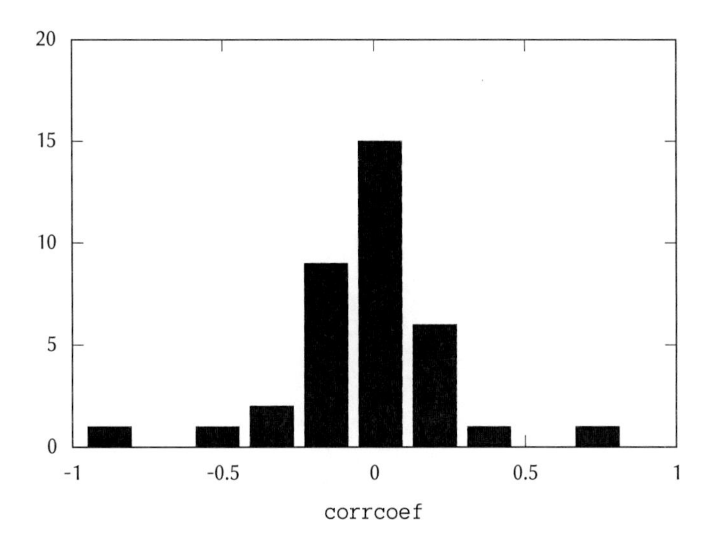

Figure 5.6.: PKD model, Listing B.2.The histogram of the upper triangular elements of the posterior sample correlation matrix excluding the diagonal elements (all 1.0). Extreme values (close to ± 1) appear twice, high correlations make sampling more challenging for simple algorithms (with Euclidean metric). This serves as a summary of Figure 5.5.

$p(\theta)$	model	size		Riemannian Manifold: RMHMC	Riemannian Manifold / Newton-Raphson: NR+RMHMC	Newton-Raphson: NR+HMC	HMC
u	MAPK	2×2	v in s^{-1}	0.312(15)	0.811(40)	4.19(29)	0.682(88)
			$v^{\text{rel.}}_{\text{eff.},\ell}$	1	2.60(25)	13(2)	2.19(39)
			$\tau_{\text{int.},\ell}$	1.491(70)	1.633(80)	3.60(24)	16(2)
u	Mma	3×6	v in s^{-1}	$0.58(10) \times 10^{-2}$	$1.35(24) \times 10^{-2}$	0.376(65)	$2.46(86) \times 10^{-2}$
			$v^{\text{rel.}}_{\text{eff.},\ell}$	1	23(8)	640(230)	42(22)
			$\tau_{\text{int.},\ell}$	18(3)	19(3)	17(3)	42(15)
u	Mifa	6×14	v in s^{-1}	$< 9 \times 10^{-6}$	$1.69(25) \times 10^{-2}$	$4.92(75) \times 10^{-1}$	$0.52(13) \times 10^{-2}$
			$v^{\text{rel.}}_{\text{eff.},\ell}$	1	$> 1930(290)$	$> 55(8) \times 10^3$	$> 58(15) \times 10$
			$\tau_{\text{int.},\ell}$	NA	11(2)	12(2)	42(10)
i	Mifa	6×14	v in s^{-1}	NA	$2.36(12) \times 10^{-1}$	$1.76(41) \times 10^{-1}$	$0.45(13) \times 10^{-2}$
			$v^{\text{rel.}}_{\text{eff.},\ell}$		52(18)	39(20)	1
			$\tau_{\text{int.},\ell}$		0.87(4)	36(8)	62(17)

Table 5.4.: Performance analysis for the original RMHMC for ODE models and two steady state data adapted HMC algorithms. The problem size is $n \times m$, where n is the number of state variables and m the number of parameters. The properties listed are the effective sampling speed v and relative speed $v^{\text{rel.}}_{\text{eff.},\ell}$) and the integrated auto-correlation length $\tau_{\text{int.},\ell}$. HMC with flat metric (last column) performs less efficient moves through parameter space, which sometimes results in higher auto-correlation. Though in our examples this efficiency loss is often compensated by lower computational costs. The last example, where we resampled the same 6×14 model with an informative prior on half of the parameters, illustrates the advantages of RMHMC over HMC. The models are listed in Appendix B.

problem size		NR+SMMALA	SMMALA
2×2	$v_{\text{eff.},\ell}$ in s^{-1}	90(2)	71(2)
	$\tau_{\text{int.},\ell}$	0.88(2)	0.88(2)
3×6	$v_{\text{eff.},\ell}$ in s^{-1}	62(1)	32(1)
	$\tau_{\text{int.},\ell}$	1.20(3)	1.52(4)
6×14	$v_{\text{eff.},\ell}$ in s^{-1}	0.30(7)	0.031(6)
	$\tau_{\text{int.},\ell}$	290(67)	166(31)

Table 5.5.: Effective sampling speed measurements for SMMALA and the modified NR+SMMALA. Both algorithms were implemented in MATLAB for these tests. The model handling and numerical ODE integration in SMMALA was done using the SBPOP toolbox (Schmidt and Jirstrand 2006). The simulations were done on a desktop computer with 12 cores.

scaled down by the step size parameter $s(m)$:

$$\begin{aligned} \phi &\sim \mathcal{N}(\theta_i, C_i^{\text{am}})\,, \\ \text{where } C_i^{\text{am}} &\approx s(m) * \operatorname{cov}(\Theta_{j<i})\,. \end{aligned} \tag{5.23}$$

This method exhibits diminishing adaptation: C^{am} converges during sampling. It is easy to implement and very fast for probability densities that resemble Gaussians. Yet, sometimes this is not a good approximation of the target distributions local shape. The adaptation of HMC on the other hand is dictated by the log-PDF's gradient, which is appropriate and informative everywhere. To illustrate this, we have designed a target distribution that isn't comparable to a Gaussian, so that its covariance is not informative at all:

$$\begin{aligned} V(\theta) &= \beta(\|\theta\| - 2)^2\,, \\ p(\theta) &\propto e^{-V(\theta)}\,, \\ \nabla_\theta V(\theta) &= 2\beta(\|\theta\| - 2)\theta\,. \end{aligned} \tag{5.24}$$

We can adjust this target using β, a free parameter, similar to a thermodynamic (inverse) temperature. It can be used to make the target density slimmer (more difficult to sample). This target density is not based on Bayesian model analysis, instead it is given explicitly in terms of θ. We have reimplemented AM and HMC for this simplified set-up. Figure 5.7 shows the

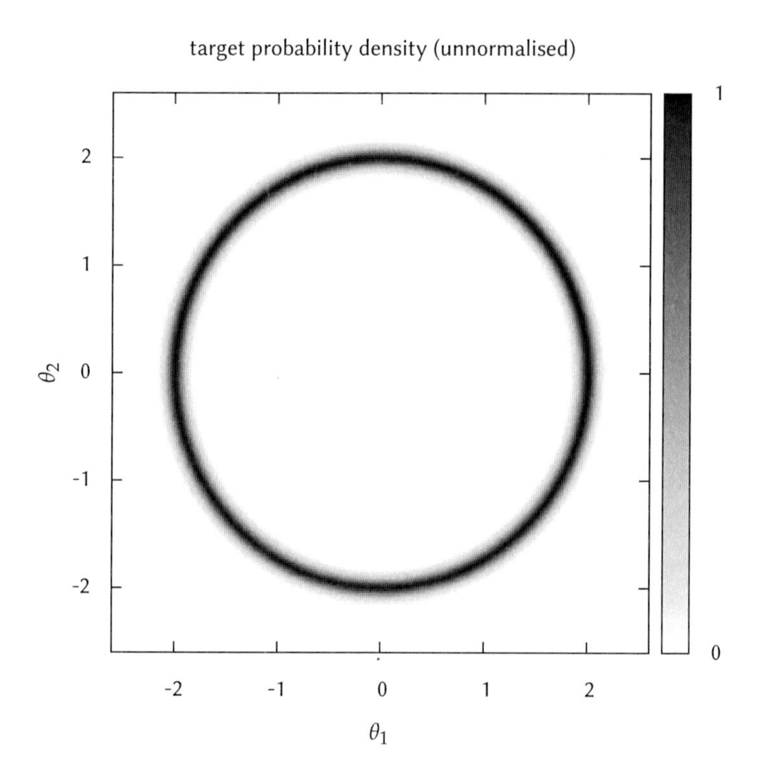

Figure 5.7.: Target probability density function for the comparison between adaptive metropolis and Hamiltonian Monte Carlo. Dark shading represents high probability density.

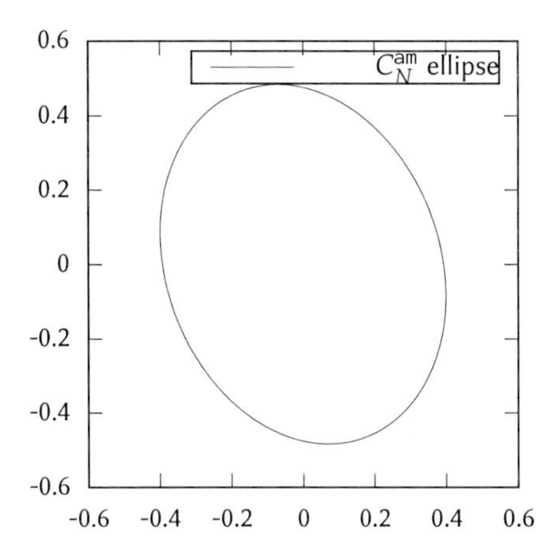

Figure 5.8.: Final shape of the transition kernel, of adaptive Metropolis, characterised by C^{am}, illustrated using an ellipse.

shape of the target PDF. Although AM remains fast, its adaptation strategy is ineffective, while HMC is entirely unhindered. The final shape of AM's transition kernel is shown in Figure 5.8 and Figure 5.9 shows the sampling results (scatterplot and estimated densities). We have used a kernel density estimate with manually adjusted kernel size (Parzen 1962; Rosenblatt 1956).

Algorithms from the Langevin family are very similar to HMC in the regard that they *adapt* to the target distribution using the log-posterior's *gradient* and possibly the Fisher information as the *metric tensor* (SMMALA). Since the shape of the a-posteriori distribution is unknown, we consider it prudent to use a locally adapting algorithm such as SMMALA or RMHMC. Nonetheless, adaptive Metropolis performs very well in most cases and densities that don't have a clear mode are rare in practice. It is especially effective when combined with delayed rejection (DRAM) and will outperform HMC because of reduced computational costs in high dimensional problems for Gaussian-like distributions. In our example HMC was faster and covered the overall shape of the target probability density better. The results are summarised in Table 5.6.

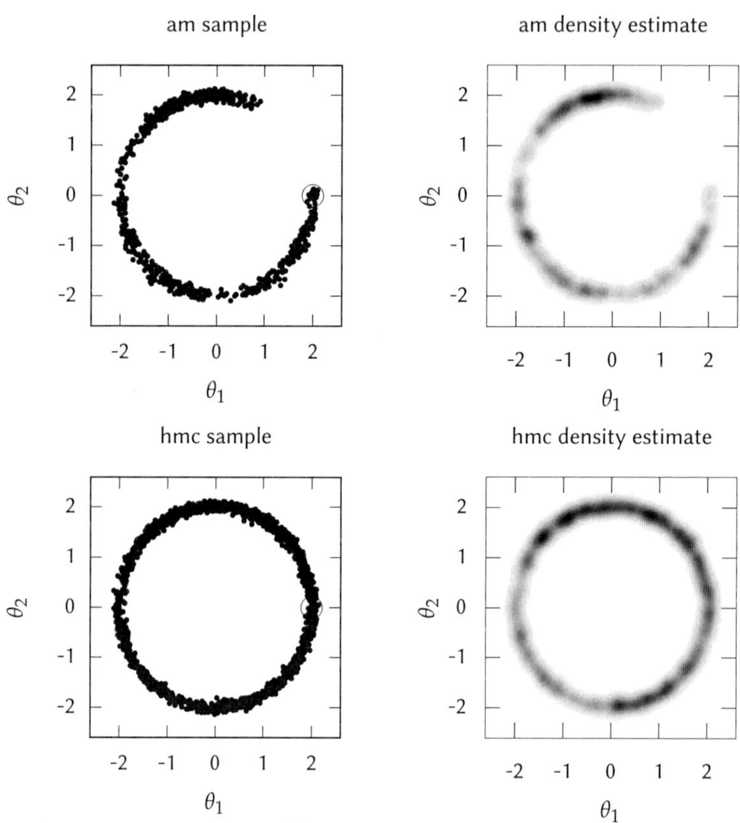

Figure 5.9.: Sample $N = 2^{11}$ and (uniform) kernel density estimate for adaptive Metropolis and HMC. The target distribution (5.24) (Figure 5.7) was not recovered well by AM in contrast to HMC. (left) sampled points, the small circle marks the Markov chain's starting point $(2, 0)$. (right) kde with uniform, Gaussian kernel.

	AM	HMC
$v_{\text{eff.},\ell}$	$1.01(16) \times 10^1$	$1.41(20) \times 10^1$
$v^{\text{rel.}}_{\text{eff.},\ell}$	1	1.40(43)
acceptance	31 %	71 %
$\tau_{\text{int.},\ell}$	3	2

Table 5.6.: Effective Speed of adaptive Metropolis and Hamiltonian Monte Carlo for target density (5.24). Sample size: $N = 2048$.

5.3.4. Initial Conditions Naturally Select Stable Branches in Bifurcation Diagrams

Using standard methods for model output calculation (ODE solvers) requires known initial conditions. Specific needs in a given experiment might require more elaborate setups. One such case is present in one of our examples, described in Appendix B.1. The problem size is 11×26 with inputs $u \in \mathbb{R}^4$ (see Listing B.1); the initial conditions for this model are known to be stable steady states, which unfortunately remained unquantified. For this reason we have defined the input function to be sigmoidal, a smooth step function, that activates any experimental perturbation of the model's reference behaviour[5] at $t = 0$. At the same time we have set t_0 to $-72\,\text{h}$, which gives the model 72 simulation time units to reach a stable steady state. This nominal, initial steady state is then changed by the activated input at $t = 0$ and the model reaches a new steady state, which is measured by the output function. Here, the initial conditions x_0 determined which steady state branch was picked (pre and post perturbation). No bifurcation analysis is performed to decide which branch, if there are more than one, appears most plausible for this model. Some algorithms avoid these simulations partly, e.g. NR+RMHMC, others entirely CBA+SMMALA. In case of multistable systems the mechanism of stable branch selection must change for these algorithms. We do not propose a general purpose method to alleviate this burden but want to make the reader aware of the problem. In NR+RMHMC, the algorithm tracks a stable branch and it is necessary to make sure that the branch is not lost on bifurcation points. When the branch choice is difficult, relative data becomes very hard to compare to the model, since now we have a branch choice to make for both the current experiment and the reference

[5] the control

experiment.

We want to illustrate the problem of relative measurements using the second algorithm, CBA+SMMALA. This method returns *all* steady state solutions $\bar{x}(\rho, u_k)^{\{a\}}$, $a = 1, \ldots, n_a$. This set of solutions has to be filtered for the biologically plausible steady states automatically to obtain a unique model output. We have decided to compare all possible solution outputs $h(\bar{x}(\rho, u_k)^{\{a\}}, \rho, u_k)$ with the observed data to find the best fit each time the model is evaluated using the CBA. In this sense the sampling process assumes the most favourable scenario for the model to make a fit.

Relative measurements make this filtering more difficult. As a specific example consider this output function:

$$y_k = h(\bar{x}(\rho, u_k), \bar{x}(\rho, u_0)), \tag{5.25}$$

$$y_{ik} = \frac{\sum_j C_{ij}\bar{x}_j(\rho, u_k)}{\sum_j C_{ij}\bar{x}_j(\rho, u_0)}. \tag{5.26}$$

Of course, the CBA might well return multiple solutions $\bar{x}_k^{\{l\}}$ in each case k. Let us list all combinations of stable steady states $\bar{x}_k^{\{j\}}(\theta, u_k) =: \xi_{jk}$ ($j = 1, \ldots, n_\text{r}$) and respectively ξ_{l0} ($l = 1, \ldots, n_\text{c}$) for the reference experiment. To calculate the possible outputs h we organise all steady state pairs in the following matrix:

$$H(k) = \begin{pmatrix} h(\xi_{1k}, \xi_{10}) & \cdots & h(\xi_{1k}, \xi_{n_\text{c}0}) \\ \vdots & \ddots & \vdots \\ h(\xi_{n_\text{r}k}, \xi_{10}) & \cdots & h(\xi_{n_\text{r}k}, \xi_{n_\text{c}0}) \end{pmatrix}. \tag{5.27}$$

$H(k)$ contains all possible experiment/reference experiment steady state ratios. Some of these pairs are impossible to observe in a real experiment, even if they fit the data really well. If the experiments are tied by similar initial conditions, which are not used here, we have no way to find *mismatched* pairs of outputs.

Let's consider the bifurcation diagram in Figure 5.10 for illustration (or any multi-stable system for that matter). Several state branches exist for $\rho > 0$, two of which are stable. On the left, the reference system is depicted. Given identical initial conditions for both systems, it is easy to construct a case where both systems can be in either the upper steady state or both in lower steady state but not in cross states (up-down or down-up): $\rho = 0.07$, $x_0 = 2$. In such a case the off-diagonal elements of $H(k)$ would be forbidden.

Such rules for picking branches for any conceivable parametrisation of the model would complicate the sampling algorithm tremendously. It does not follow directly how the branch selection rules change when $\rho = 0.02$; the reference experiment $u = 1.0$ has one branch, while $u = 1.1$ provides two branches.

Since there is no automated way to filter out such impossible pairs of steady states, we advise caution and restrict our own examples to absolute observations of the form $y = h(\bar{x}, \theta, u)$. Of course, it might be unambiguously clear how branches should be picked from the initial problem statement, which resolves this issue. In addition, this problem is not entirely alien to the other algorithms, but is masked whenever fixed initial conditions are used to obtain an unambiguous model response[6]. So, great care must be taken in any case.

5.4. Model Assisted Design of Experiments

Sampling methods can not only be used to fit a model to previously obtained data, but also to make predictions for unobserved scenarios which can inform the next experimental setup. We consider two different approaches for the experiment design problem (Kramer and Radde 2010; Weber, Kramer, et al. 2012); these methods can complement each other since they work with different assumptions. The *first* method is based on entropy and operates in parameter space. Prior knowledge is used to make a prediction for each proposed experiment. We then select one based on the expected gain of information. The *second* method works in the space of model outputs y and uses their prediction uncertainty to find the best measurement time.

The sampling approach to model analysis makes these methods very easy to implement and the discussion of these methods gives us an opportunity to reevaluate the importance of steady state experiments.

5.4.1. Entropy Based Selection of Experiments

The general idea of this method is to evaluate the expected information content of posteriors that can be generated through the proposed experiment. This procedure was published in Kramer and Radde (2010). It is numerically expensive, because it calculates an expectation value of information over

[6] possibly unjustifiably so

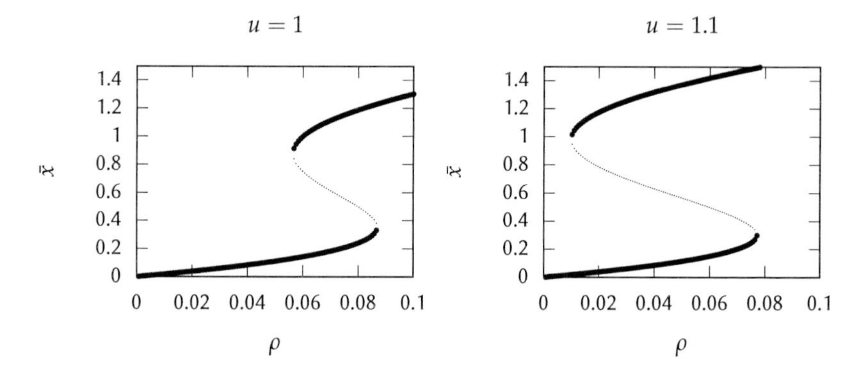

Figure 5.10.: (left) bifurcation diagram of the reference system $\dot{x} = \rho - \beta x + u_0 x^2/(1+x^2)$ with $u_0 = 1$; (right) bifurcation diagram for $u = 1.1$. The output function is of the form (2.7). This was computed using the `roots` function, *thick lines* represent stable branches, *thin, dotted* lines represent unstable branches. This demonstrates a general problem for MCMC algorithms for relative data. For some parameter choices near $\rho \approx 0.07$ it is hard to decide which two stable steady states (one from the left and one from the right frame) to pick to compute the output function h. If some selection rules are known to the experimentalist (e.g. always pick the upper value) it is not necessarily obvious how to apply the rule when the parameter ρ is changed. (we have used: $\beta = 0.56$)

the space of possible experiment outcomes (which are uncertain in our framework).

An important notion in this procedure is that it iterates, so a current posterior and its attached sample can be viewed as the prior for the next iteration, which includes sampling every time. On each iteration we try to select the most promising experiment from a list of descriptions: $\mathcal{E}_k$, $k = 1, \dots, n_E$. This list can change on each iteration. Of course, the Bayesian framework can process $\mathcal{D}$ from more than one experiment, but for our purposes we assume that exactly one experiment is performed next; this decision merely has consequences for notation. Next, we describe the basic structure of one iteration.

Greek letter H, η (Eta)

One of the most interesting sample summaries with regard to information of a sample is the Shannon entropy $H_S(\mathcal{D}, M)$, which in our case reads as

$$\begin{aligned} H_S(\mathcal{D}, M) &= -\int p(\theta|\mathcal{D}, M) \log p(\theta|\mathcal{D}, M) d\theta\,, \\ &= -\langle \ell \rangle\,. \end{aligned} \tag{5.28}$$

We can use the entropy to select the experiment $\hat{\mathcal{E}}$ with highest information content from the list (Kramer and Radde 2010). Since the data of a proposed experiment is not known yet, it is predicted using the model M by drawing a series of representative parameter values θ_l (ρ_l, $l = 1, \dots, n_{\text{ED}}$) from the currently available prior and simulating the model. Each θ_l implies an output $y_{jkl} = h(x_{jk}, x_{j0}, \rho_l, u_k)$, which we use as virtual data $\mathcal{D}_{kl}$ for MCMC sampling. This has to be repeated for each experiment $\mathcal{E}_k$, with its own input u_k and list of measurement times t_j. This list may be different for each experiment, although we do not reflect this fact in our notation.

During sampling, k and l have an influence on the virtual data only, which is fixed for each MCMC run. Most crucially, k and l are never *summed over* within the likelihood function. Both k and l always appear on both sides of equations during Bayesian analysis; to simplify notation, we colour them *gray*: $y_{ijkl} \in \mathbb{R}$.

Since k and l are nested iterations, the sampling is repeated $n_E n_{\text{ED}}$ times. Each MCMC run yields a *predicted* sample Θ_{kl}: it is predicted for a given experiment $\mathcal{E}_k$ and hypothetical model parametrisation θ_l ($l = 1, \dots, n_{\text{ED}}$). In aggregate, all n_{ED} parameter vectors in Θ_{kl} represent the current state of knowledge available for model based predictions. Next, we summarise each sample Θ_{kl} by its entropy $H_S(\mathcal{D}_{kl}, M)$, to be able to merge the results by averaging over l. We use each sample (with sample index r) to estimate the entropy H_S:

$$[H_S(\mathcal{D}_{kl}, M); N] = -\frac{1}{N}\sum_{r=1}^{N} \log p(\theta_r|\mathcal{D}_{kl}, M), \tag{5.29}$$

$$=: -\frac{1}{N}\sum_{r=1}^{N} \log \pi_{rkl}, \tag{5.30}$$

where, unfortunately, π_{rkl} is usually unknown, due to unknown evidence $p(\mathcal{D}_{kl}|M)$ for the model. To solve this problem, we estimate the evidence as well. Many approaches exist, one very good technique is described in Calderhead and Girolami (2009). For smaller problems a naïve approach works rather well: since we do have the unnormalised posterior values w_{rkl} (for any fixed l and k)

$$w_{rkl} = \exp\left(-\frac{1}{2}\sum_{ij}\left(\frac{y_{ijkl} - h_i(x_{jk}, x_{j0}, \rho_r, u_{kl})}{\sigma_{ijkl}}\right)^2 - \log(p(\theta_r|M))\right). \tag{5.31}$$

The parameter σ_{ijkl} should be a reasonable error estimate derived from experience with the proposed experimental setup, it is the predicted noise. We determine the normalising factor using kernel density estimates (kdes). Because kdes are normalised, the constant normalising factor ν_{kl} can be inferred by comparing w_{rkl} to the kde values (Kramer, Hasenauer, et al. 2010). Figure 5.11 illustrates this procedure, for fixed, omitted k and l. The normalisation method does not require the full sample. But, using subsamples of increasing size provides increasingly better estimates; this is illustrated in Figure 5.12.

Once we have found the normalising factor $\hat{\nu}_{kl}$, we calculate the l-average expected entropy $\eta_k(M)$ for each experiment:

$$\mathcal{E}_k \mapsto \eta_k(M) = -\frac{1}{n_{\mathrm{ED}}}\frac{1}{N}\sum_{l}\sum_{r=1}^{N} \log\left(\frac{w_{rkl}}{\hat{\nu}_{kl}}\right). \tag{5.32}$$

The most informative experiment, the one with the highest η_k, for the used model M, is chosen to be performed on the real system. The results of this real experiment may be processed using the Bayesian framework. The resulting posterior may then be used if the experiment design is iterated. Due to this method's high numerical costs of $n_E n_{\mathrm{ED}}$ repetitions of MCMC sampling, it is not suitable to optimise real valued parameters of an experiment, such

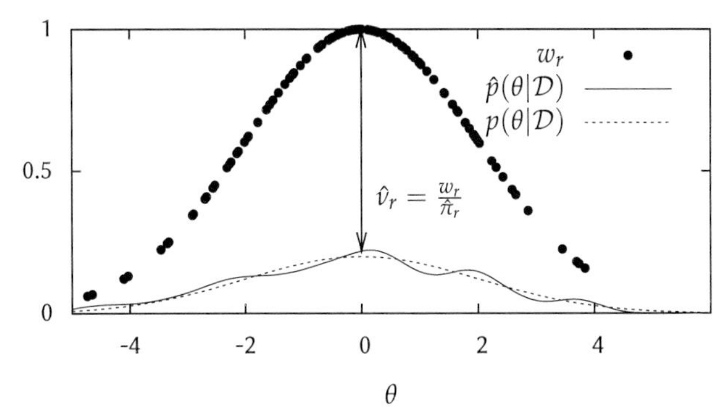

Figure 5.11.: Illustration of the normalisation method for a Gaussian distribution with one degree of freedom and sample size N. We propose to use the unnormalised sample of posterior distribution values $\{w_r\}_1^N$ and compare them to the kernel density estimator values $\hat{\pi}_r = \hat{p}(\theta_r|\mathcal{D})$ to find the normalising factor $\hat{v} = [\hat{v}_r] = [w_r/\hat{\pi}_r]$. In this manner we obtain normalised target density values for the sample. Symbol summary: $\pi_r = p(\theta_r|\mathcal{D})$ *(true posterior)*; $\hat{\pi}_r = \hat{p}(\theta_r|\mathcal{D})$ *(kernel density estimate)*. (Kramer, Hasenauer, et al. 2010)

as measurement times or input values. If measurement times $t_j \in \mathbb{R}$ have to be optimised we can use the method described in the following Section.

5.4.2. Variance Based Design of Experiments

An altogether different approach from the one described in Section 5.4.1 is to use the model trajectories derived from the MCMC sample to find a time point $\hat{t}$ at which the largest model output uncertainty occurs unless the prediction uncertainty is lower than the expected measurement error. If no time-point t (within the available window) yields a predicted trajectory variance above measurement uncertainty, then this design method terminates. The next series of experiments should record measurements as close to $\hat{t}$ as possible (Weber, Kramer, et al. 2012).

Figure 5.13 and 5.14 demonstrate this procedure for the PKD model (Listing B.2) using a proposed experiment with $u_1 = \pi$ and $u_2 = 2\pi$ (chosen arbitrarily for this demonstration). It seems prudent measure one (or all) of

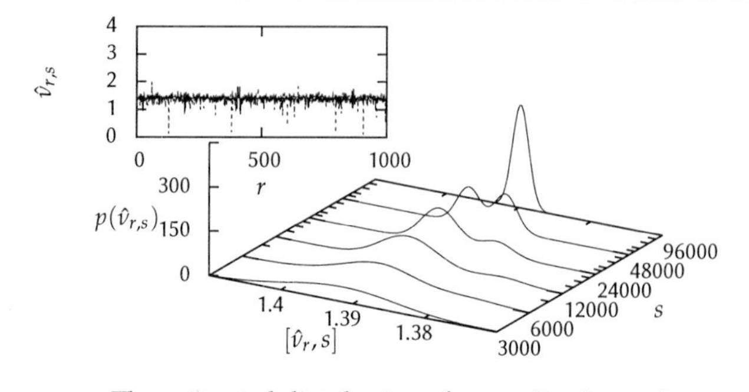

Figure 5.12.: The estimated distribution of normalisation estimates ν derived from subsamples of increasing size s. A large subsample reveals an accurate estimate of the normlaisation factor. The small box shows an example of individual normalisation factors $\nu_r = w_r / \hat{\pi}_r$, for a small subsample of size $s = 1000$. In the small box, the solid line shows values using underlying parameters θ_r sampled directly from the posterior, while the dashed line uses parameters sampled from the density estimator (given the whole sample). This is to check the accuracy of the kernel density estimate. Both methods agree. (Kramer, Hasenauer, et al. 2010)

the previously neglected quantities:

$$y_{ijk} = h(x(t_j; \theta, u_k))$$
$$h(x) = \begin{pmatrix} x_2 \\ x_4 \\ x_6 \end{pmatrix} . \tag{5.33}$$

Here, t_j is subject to optimisation. We have covered a predefined range $t \in [0, 11]$ with a dense array of possible t_j ($j = 1 \ldots, T$, $T \gg 1$) and plotted the predicted output trajectories.

The highest uncertainty is predicted for the steady state of x_2, since y_1 has the highest standard deviation in Figure 5.14. But, we also note that *all* considered observables x_2, x_4 and x_6 are more uncertain in this prediction than the available measurement accuracy (dashed line in Figure 5.14). It is therefore useful to measure all three steady states in the next series of experiments. We emphasise that the proposed measurement times are steady states; $\hat{t}$ is indicated by filled dots in Figure 5.14. Visual inspection of $\hat{t}$ in Figure 5.13 shows that the trajectories are very flat: $\hat{t}$ lies in the steady state regime of this model. This outcome is very intuitive: the output trajectories approach the steady state in a monotone fashion, so the differently parametrised trajectories have had the *most* time to disperse *precisely* when they have reached steady state. This property renders the steady states the most uncertain feature of the model. Consequently, accurately measured steady states are more useful than a recorded time series. Of course, these conclusions rely heavily on the accuracy of the used model. The true results of the experiment, once performed, might be very different from these predictions.

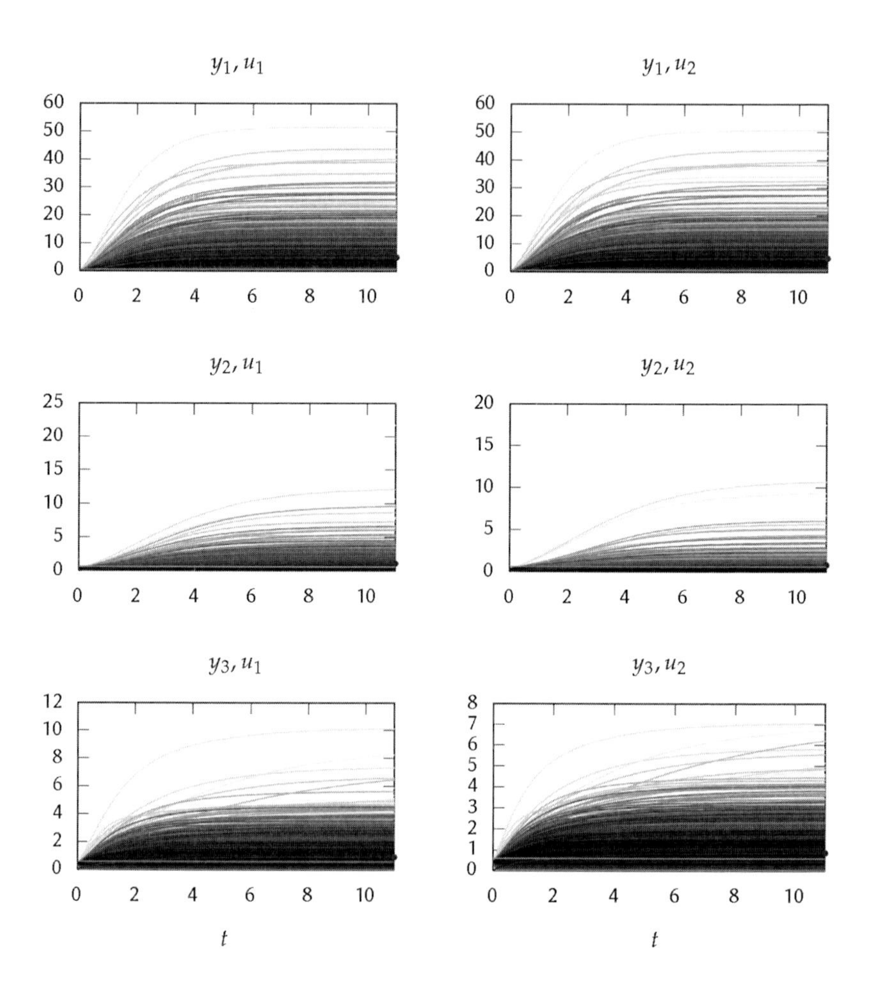

Figure 5.13.: Trajectories of the PDK model (Listing B.2) using the posterior parameter sample obtained via SSMMALA, depicted in Figure 5.4. The proposed experiment has the output function $h(x, \rho, u_k) = (x_2, x_4, x_6)^{\mathrm{T}}$ and the inputs are (arbitrarily) set to $u_k = k\pi$. Shown are the trajectories in grayscale (depending on ℓ), the red line shows the standard deviation of the trajectory bundle, the blue line shows the measurement uncertainty of the previous experiment (the expected accuracy of the planned experiment).

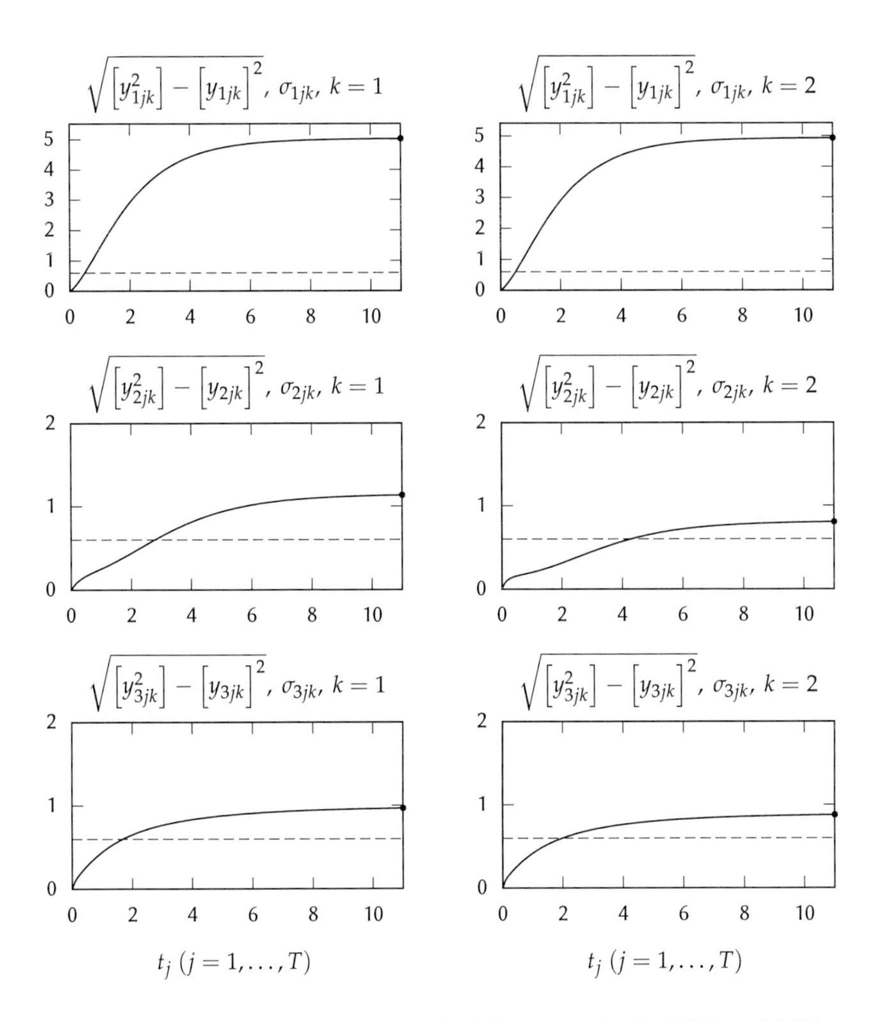

Figure 5.14.: Trajectory bundle standard deviations for the PDK model (Listing B.2) using the posterior parameter sample obtained via SSMMALA. The proposed experiment has the output function $h(x, \rho, u_k) = (x_2, x_4, x_6)^{\mathrm{T}}$; the inputs are set to $u_k = k\pi$. The dashed lines show the measurement uncertainty of the previous experiment (the expected accuracy of the planned experiment).

6. Software Development

The development of MCMC software is rather difficult because the measurement model can be rather complex and must be reflected by the likelihood function. In a piece of software the numerical integration of the model must be performed to calculate the value of the output function h, depending on the structure of this function, the integration has to be performed several times under different conditions. Because of this, high level script languages such as MATLAB are very convenient, because the code can be changed by the user to accommodate any special likelihood function. On the other hand, sampling is very time consuming and it is very desirable to have a sampling software written in a speed oriented language such as `C`. The software package `mcmc_clib` is the result of our endeavour to fill this need, for at least some common measurement cases in systems biology.

6.1. `mcmc_clib` – Software Package for Bayesian Model Analysis

This software is intended for very big models, with a complex interaction structure (many feedback and feed forward loops) and time series data. The algorithm used is the SMMALA algorithm with numerical ODE integration performed by the CVODES module of the SUNDIALS solver suite. All necessary analytical computation is done by VFGEN via GiNaC (Bauer, Frink, and Kreckel 2002; Weckesser 2008).

The general output model of `mcmc_clib` supports relative data of the form (2.7). The model setup language makes a distinction between a primary output structure as described in (2.10) and the fully quantitative secondary layer of the output function (for relative data). The model and experiment structure are defined in two different files, the model file contains the primary output function $g(x_{jk})$ while the experiment file determines whether the output is relative to a reference or not. In the former case the full output function is $y_{jk} = h(x_{jk}, x_{j0}) = g(x_{jk})/g(x_{j0})$. This layered approach is one of the strengths of this package and very convenient for systems biology

implementation	SMMALA gcc compiled C	SMMALA GNU OCTAVE	DRM GNU OCTAVE
$v_{\text{eff.},\ell}$ in s^{-1}	0.842(40)	$2.22(58) \times 10^{-4}$	$1.19(18) \times 10^{-3}$
$v^{\text{rel.}}_{\text{eff.},\ell}$	$3.8(1.0) \times 10^3$	1	5(1)
acceptance	53 %	54 %	28 %
$\tau_{\text{int.},\ell}$	83(4)	83(22)	166(25)
sample size N	2^{20}	2^{14}	2^{17}
CPU time t_s	2.09 h	124 h	92 h
wall clock time	2.09 h	15.7 h	18 h
step size ϵ	0.1847	0.1847	0.15

Table 6.1.: A comparison of SMMALA implemented in C and GNU OCTAVE sampling the parameters of the PKD model. The implementation of mcmc_clib is not parallel, i.e. it does not benefit from the multicore cpu, therefore wall clock time is identical to CPU time. For comparison, we have also included results from DRM tests (see Section 3.5.4).

applications.

We have used the software package to repeat the tests in Section 5.3. Since this implementation is very fast, we have chosen a far larger sample size with $N = 2^{20}$, it is displayed in Figure 6.1. Once again we have obtained samples for the PKD model (size 6×9) and measured sampling time t_s. We compared two different implementations of standard SMMALA (in C and GNU OCTAVE) and provide test results for the same problem for the delayed rejection Metropolis algorithm. The model is described in Appendix B.2, Listing B.2; once again we have used simulated data obtained through forward integration of the model and measurement noise using random numbers (randn() function). Table 6.1 shows the effective sampling speeds of all three sampling procedures along with some other characteristic measurements such as auto-correlation time $\tau_{\text{int.},\ell}$. For this model size, mcmc_clib proves to be vastly superior to both OCTAVE implementations.

6.2. mcmc_clib – Usage Structure

Here we summarise the software as an input/output system with:

Input The ODE model has to be provided as an xml file in VFGENS vf

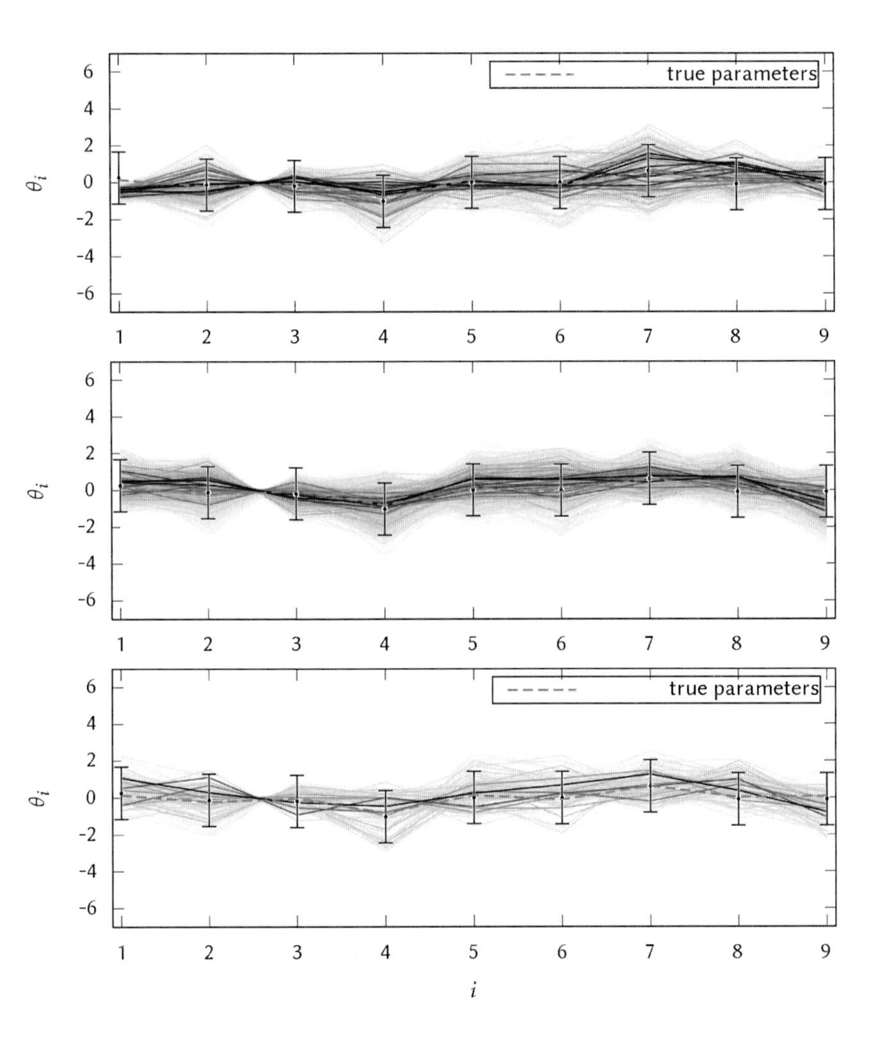

Figure 6.1.: The posterior sample for the medium sized (6 × 9) PKD model and time series data. From top to bottom: mcmc_clib, DRM, SMMALA. The latter were both implemented in GNU OCTAVE. Once again only the effective sample size (4.3) is shown.

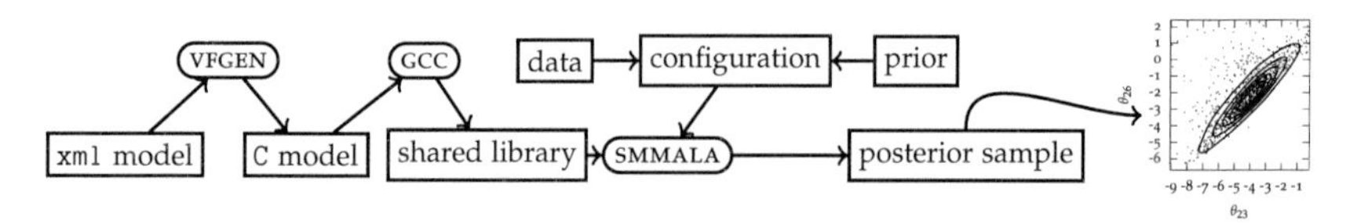

Figure 6.2.: Flowchart of the software's usage structure (Kramer, Stathopoulos, et al. 2014). The model is conveniently stored as a VFGEN vector field file. VFGEN generates C source files for: the vector field, its Jacobian, the output functions and their sensitivities. These sources are compiled to a shared library which can be loaded by the smmala sampler. The sampler also receives settings, such as step sizes, desired sample size but also experiment conditions (inputs) and data in the configuration file. With these inputs the sampler generates a sample which is written into an output file.

format, together with a data file that contains measurement time points and a data set $\mathcal{D}$. Additionally, this data file also specifies sample size, initial step size, desired acceptance rate and the parameters μ and Ξ^{-1} for the prior distribution.

Sampling algorithm To evaluate the likelihood function ODEs are numerically integrated using CVODES with fixed relative and absolute tolerance. The SMMALA method is used to produce proposals ϕ.

Output The user can select a text or binary file output, which contains the results of the sampling procedure: the sampled log-parameters θ_i and their log-posterior values ℓ_i. The sample can be further processed with other numerical software (e.g. GNU OCTAVE, MATLAB).

This is illustrated in the flowchart, Figure 6.2.

6.3. mcmc_clib Is Capable of Dealing with Large Models

The mcmc_clib software is designed to cope with large scale systems; an appropriate model with 11 state variables and 26 parameters is packaged with the software for testing purposes. The data for this example stems from a real experiment but model and data are anonymised.

We have published an application note introducing this package (Kramer, Stathopoulos, et al. 2014). Listing B.1 shows the vf file format which we use to store the example model. It differs from the sbml format, which is widely used in systems biology, to make the model import more manageable.

Specifically, sbml offers the option to define measurement units, even *different units* for different variables of the *same type* (e.g. concentrations). This, among other properties, makes the systems biology markup language more powerful as a modelling language but also more complicated for the task of defining an ODE model for *numerical simulation*. For post processing purposes or to make predictions using the sample, the model has to be exported into various different languages such as R, MATLAB or GNU OCTAVE. The vf format is very well suited for these purposes and the cross conversion is already implemented in the VFGEN software for most languages including *all of the above* and C.

To accommodate the concept of *relative measurement* we use a two layered definition. The primary output function is defined in the model file. This primary output simply defines which quantities are measured (in arbitrary units). A second file defines the experimental setup and asserts whether the data is relative. Only in that case, a reference experiment and data set are defined. Then, the main program ode_smmala itself will perform the ratios between measurements and reference measurements. The ratio (2.7) forms the second layer of the output function and is done *once* for the data set but also repeatedly for the *model output* h for each parameter vector during sampling. A summary of this comparison is listed in Table 6.2. Note that the model definition in Listing B.1 only includes the first layer of the output function; the primary output variables are named Y1 to Y6; from this definition alone, we cannot yet tell, whether the measurements are absolute or relative. This is determined by the data file supplied to the software, specifically by the occurrence of the tag [reference_data] or lack thereof.

To demonstrate the difference between the two setups we have written an example data file with relative measurements and an additional file where we include the reference data alongside the rest in one data matrix and pretend that the numbers are absolute measurements (they are not). The amount of data points remains the same for both setups, albeit with different interpretation. We have drawn very different samples for these setups, shown in Figure 6.3. The features are summarised in Table 6.2. It was more difficult for us to adjust the algorithm to the case of absolute data (it is harder to reproduce with the model), this is somewhat reflected in the values shown in Table 6.2.

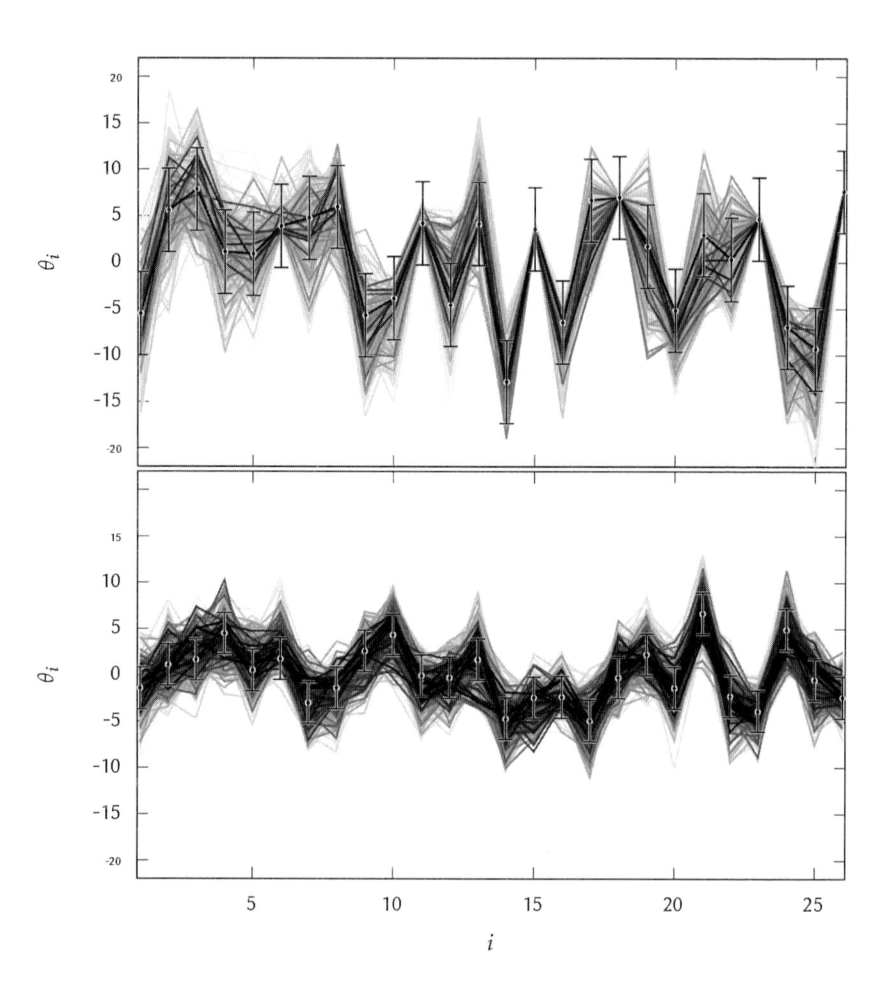

Figure 6.3.: *(top)* posterior sample of the model in Listing B.1 using the measurements as absolute data. *(bottom)* posterior sample of the same model but data correctly treated as relative measurements. The two samples are very different, in both cases they are colour coded using the unnormalised posterior density values; darker lines represent higher values. Both priors were obtained from initial sampling with fairly uninformative priors (Gaussians with $\sigma \gg 1$).

`mcmc_clib`	absolute data	relative data
sample size N	2^{18}	2^{17}
effective sampling speed $v_{\text{eff.},\ell}$	$1.43(34) \times 10^{-3}$	$0.82(15) \times 10^{-2}$
acceptance a	17 %	42 %
auto-correlation $\tau_{\text{int.},\ell}$	$1.01(24) \times 10^{3}$	$0.260(48) \times 10^{3}$
step size ϵ	8×10^{-3}	0.15149
sampling time t_s	25 h	8 h

Table 6.2.: Sample Properties for the model in Listing B.1. The data used in this test is indeed relative and a reference experiment is provided, treating the data as if it was absolute is nevertheless possible, but a completely different problem. The posterior parameter samples differ substantially for these two cases.

7. Conclusions

We have emphasised the usefulness of ordinary differential equations for systems biology and laid out how they relate to other models in Chapter 2. We set up our modelling framework to benefit from known steady state sensitivities and their relation to the parameter space metric. In Section 2.1, we have motivated the development of our algorithm variants [1] with the notion that they should adapt to the type of data that is available. Finally, we have laid out how measurement uncertainty ties the models to a Bayesian parameter estimation framework and probabilistic predictions.

In Chapter 3, we have introduced three general types of sampling algorithms suitable for ODE models: Metropolis-type, Hamilton-type and Langevin-type methods. We have derived algorithm variants which use more of the problems mathematical structure in Chapter 5. Specifically, these methods use derivatives of the ODE's right hand side function f and take the knowledge about how the observed data was obtained into account.

The use of the metric tensor $G(\theta)$ in Riemannian manifold methods allows moves through parameter space that are more efficient, because they take some of the posterior's shape and curvature into account. It is calculated using the model's output sensitivity.

We have used an approximation for the time dependent sensitivity of $x(t, \rho, u)$. These approximations require $x(t, \rho, u)$ to approach a stable steady state for large t. The approximation is exact for steady state data. The much lower computation time of this sensitivity approximation leads to a reduced sampling time t_S, which is shown in Table 5.3. In addition, we can use it to calculate the approximate metric tensor.

It is certainly possible for our approximation of $G(\theta)$ to perform worse than the exact matrix in its effect on the sampling speed, because it may estimate the posterior's curvature badly. This effect did not outweigh the benefits in any of the examples we have encountered. We have observed an order of magnitude in speedup for our GNU OCTAVE implementation of the SSMMALA over the default SMMALA for the PKD model (Listing B.2).

[1] CBA+SMMALA, SSMMALA, NR+RMHMC, and NR+HMC

We have also tested the method of circuit breaking for the calculation of steady states. Before the sampling is initiated, we check whether the *circuit characteristic*, that characterises the steady states, is a parametrised polynomial. This polynomial is sufficient to calculate the model response during sampling, the CBA is performed only once. The construction of the companion matrix for the circuit characteristic renders the steady state problem equivalent to the calculation of eigenvalues (Section 5.1.3). The numerical computation of eigenvalues is extremely fast; the QR-method (Francis 1961, 1962; Kublanovskaya 1963) with Householder preconditioning and similar (Bates and Watts 2007), highly efficient implementations are available for these computations in all programming languages mentioned in this manuscript. Table 5.3 shows that this method improves the sampling speed dramatically; in our example two orders of magnitude for a relatively small model with $n = 6$ state variables (Listing B.2).

A larger study in MATLAB deals with the benefits of the very efficient and exact analytical steady state sensitivity analysis. This is summarised in Table 5.4. Here, we modified Hamilton-type algorithms such as RMHMC, described in Section 3.6. We have improved the RMHMC algorithm in three ways:

- Since the data has the steady state property, we can use the Newton-Raphson method to solve $f(\bar{x}, \rho, u) = 0$ for $\bar{x}$. Given a good starting point, this iterative method is quicker than a full solution to the initial value problem $\dot{x} = f(x, \rho, u)$ $(x(t_0 = 0))$.

- Taking the nature of the Hamiltonian approach into account, we can expect each step of the update trajectory calculation (3.44) to change the parameters θ only very slightly. Using steady state sensitivity analysis, we can predict the changes in steady states for small parameter changes and, by doing so, provide a very good starting point for the Newton-Raphson method.

- The exact sensitivity analysis of steady states is performed by solving linear systems of equations (3.49) and (3.51) rather than by solving initial value problems for states and sensitivities with (second order) sensitivity analysis. Solving linear systems is a lot faster than the numerical integration of ODEs and can be used cheaply at each step during trajectory integration.

Consequently, the costs of obtaining the metric tensor $G(\theta)$ become almost negligible. This means that all theoretical benefits of RMHMC over

Metropolis-type methods come at very low additional cost. This renders RMHMC effectively very competitive even for ODE models. This is a very encouraging result as the RMHMC method was shown to outperform many other algorithms for more analytically accessible models (Girolami and Calderhead 2011). All of the listed benefits are interconnected and mutually beneficial. The scaling of Hamilton-type algorithms improves so greatly that even for moderately sized models with 14 parameters we observe speedups of three orders of magnitude compared to the unmodified method. This is listed in Table 5.4. Of the methods we have compared in this manuscript RMHMC demands the highest number of output recalculations per sampling iteration and therefore benefits the most from these changes. In the case of SMMALA the speedup is comparatively moderate but well worth the effort of implementation. In our examples we observe a speedup of one order of magnitude for the Mifa model (Brännmark et al. 2010) of Insulin Dose response with 14 parameters.

In summary, we have shown simple ways of improving these two classes of algorithms under various implementations up to at least one order of magnitude for the smallest models encountered in systems biology. The curse of exponential scaling of simulation time over the number of parameters in sampling techniques implies an exponential scaling of the speedup as well. Effectively, these modifications make larger models accessible for parameter estimation, models that were not feasible before.

Section 5.4.1 shows how a sample may additionally provide a model based decision for the selection of proposed Experiments. Section 5.4.2 shows how steady state measurements can be very valuable for parameter inference and further encourages our approach to improve steady state sampling. The suggested time point $\hat{t} = 11$, with maximum posterior prediction uncertainty, is well within the time invariant regime of the PKD model's dynamics (Figure 5.13). This is because in the absence of interesting transient behaviour the model trajectories under different parametrisations diverge more and more with increasing simulation time. Once the trajectories are very close to the stable steady state, their distance also becomes time invariant (at its maximum value). Figure 5.14 illustrates this behaviour, each trajectory in the shown bundle is strictly increasing with simulation time as is the standard deviation of the bundle.

In Chapter 6, we show that the benefits of an efficient implementation of SMMALA surpasses the speedup brought forth by modifications to the algorithms for intermediate problem sizes. The system in Listing B.1 with size $(n = 11, m = 26)$ is not big enough for the better scaling of the modified

n state variables and m parameters

algorithms such as SSMMALA to prevail. Using the smaller ($n = 6, m = 9$) PKD model in Listing B.2 we have found SMMALA to be slow compared to our variants in both MATLAB and OCTAVE. On *equal ground*, using the *same programming language, model and data* , SMMALA was the slower MCMC method when compared with NR+SMMALA in Table 5.5. It remains the slowest method when it is compared to SSMMALA and CBA+SMMALA, shown in Table 5.3. Yet, using two different programming languages, we show in Table 6.1 that the `C` implementation of SMMALA outperforms its GNU OCTAVE counterpart, with a tremendous speedup of $3.8(1.0) \times 10^3$ for the PKD example. By extension, we expect `mcmc_clib` to perform better than SSMMALA or NR+SMMALA as long as they are implemented in a high level script language such as GNU OCTAVE or MATLAB. This improvement, however, comes at the high cost of having to implement complex functions for data and model import as well as a steep learning curve for users who wish to modify parts of the program, but are not familiar with `C`. The workflow of modification involves the use of: the command line, Makefiles, a `C` compiler such as gcc and additional tools such as VFGEN for model storage. Furthermore, the sample must nevertheless be post processed using a scripting language such as R for statistical analysis or GNUPLOT for visual inspection (Eaton, Bateman, and Hauberg 2009; R Development Core Team 2008; Williams et al. 1986–2014).

Looking further at the PKD example, we can see that delayed rejection Metropolis worked very well, as listed in Table 6.1. Although impossible to tell beforehand (in practice), the posterior turned out to be not very difficult to sample. Most correlation coefficients between parameter pairs turned out to be small, as summarised in Figure 5.6. For this reason, we decided to compare the performance of SMMALA to an algorithm from the Metropolis family. We have implemented the non-adaptive, standard Metropolis algorithm with delayed rejection DRM (Mira 2001) in OCTAVE and obtained a speed up of $5(1)$ over SMMALA. In this specific case the calculation of the metric in SMMALA turned out to be a burden for the sampling algorithm rather than a benefit.

The implementation of delayed rejection was not easy even in a high level programming language such as OCTAVE. Both our implementation and the `mcmc_stat` code by Haario, Saksman, and Tamminen (1998) use the recursive definition of the acceptance (3.31). We expect that a combination of SMMALA with the delayed rejection paradigm would be even more difficult considering the recursive nature of the acceptance function and asymmetric, locally varying proposal densities.

Let us consider the differences between the improvements within the MCMC methods and performance boosts due to the implementation framework. Although we have not fully investigated the scaling properties of our algorithm variants, Table 5.4 strongly suggests that larger problems have increasingly larger speedups of NR+RMHMC over standard RMHMC. This is probably not the case for simply choosing a different programming language. Although we did not analyse how the speedup of C over OCTAVE scales with problem size at all, we expect a flat benefit. Still, further investigation of the scaling might be insightful.

Nonetheless, we can conclude that the effort to develop the C version of SMMALA far outweighs the benefits of a superior algorithm at least for this problem size.

We conclude that further work on the C implementation of the SMMALA family of sampling algorithms might result in a very efficient tool, which in time might grow to be more robust in the face of possible usage errors and provide a set of problem specific samplers such as SSMMALA. The increase in computation speed seems worth the additional effort.

Outlook

Using problems of increasing size, we could test the scaling properties of `mcmc_clib` and find its current limits on desktop machines. Additionally, this implementation is well suited for MPI support to benefit from distributed computing clusters.

To improve the user experience, it would be great to transform the sampler into a function that can be called within OCTAVE, R or PYTHON. Although very time consuming, as large parts of the software have to be rewritten, the resulting method would be easy to use and very fast, benefiting from both worlds. Users could store the data and experiment setup in these high level programming environments as well as organise the overall model analysis using only the tools they are accustomed to, as R and MATLAB are very common tools in the field of systems biology. GNU OCTAVE is a free GPL licensed software that is for the most part compatible to MATLAB. So, using such a function, sampling and post-processing (plotting, predictions) would be done within the same environment.

The `mcmc_clib` software could become even more powerful if we were to add the capabilities of approximate sensitivity analysis, described in Section 5.2, or the ability to perform the Newton-Raphson method if the data is known to be in steady state.

Because mcmc_clib currently only supports Gaussian noise and a very rigid output paradigm of data and reference_data with *identical measurement times* it is expedient to include more variability within the likelihood function. A very frequently encountered, important, but currently unavailable setup, is the measurement of a *relative time series*. The recorded values are arbitrarily scaled as a group. The ratios between them are reproducible, while the absolute values are not. Defining the most accurately measured point in the series as a reference value is a good way to deal with this type of data. Currently, the reference data points *need*cto be measured at the same time points as the data they aim to normalise. So far, only one measurement time instance list t_j is defined in the data file and adhered to during simulations.

For relative time series data, the input often remains unchanged between the reference point and the rest of the data as they lie within the same time series. This has to be reflected in the likelihood function definition and cannot be done using the current syntax.

A second frequently needed function of the likelihood term is the ability to log-transform data and model outputs since a log-normal error model is very useful in systems biology.

It is hard to catalogue all standard scenarios of data post-processing. This is the main reason why high level languages are so very popular. They allow for these different scenarios by having simple enough syntax, so that users can learn to change the likelihood function according to their needs. Modelling real world systems and processing data while continuing development of the software seems like the most practical way to increase the versatility of mcmc_clib.

Nevertheless, mcmc_clib in its current state does already cover many important scenarios in systems biology and is very useful.

A. Modelling in Systems Biology

One of the first questions one has to ponder when modelling complex systems surely has to be *what is the right balance between accuracy, precision and approximation* to get sufficient predictive power and insight. And consequently, *which type of model is most appropriate for a given observation or data in terms of the state variables*. Here, we give a quick overview on the scales and granularity of available models for systems biology.

A.1. Scale and Granularity

The purpose of mathematical models is to gain understanding about the system and perhaps to a larger extent to make predictions for previously unobserved conditions. The typical questions in the field of systems biology involve the inner processes of living organisms such as bacteria, single cell organisms, eukaryotic cell cultures, tissues in a larger organism or even complex substructures of living organisms (coupled systems). Given a problem, e.g. understanding cell differentiation from generic to specialised cell types, we could try to understand the mechanism by compiling everything we know about the inner workings of cells. The top down view of this would involve the realisation that cells are subdivided into compartments (organelles) separated by membranes built from lipid bilayers. These compartments contain water, more lipids, proteins, sugars, salts and acids such as {m,t,r}RNA and of course the DNA with genes coding for specific phenotypic effects. These molecules form complex interacting systems (reaction networks) governed by the laws of motion, thermodynamics, statistics, and quantum mechanics. The atoms within these molecules share electrons in covalent bonds and/or attract each other via electromagnetic fields, while the electrons themselves are quantum mechanical fields which interact with other such fields at stochastic points. These interactions make them appear as particles.

Approaching the problem at the most fundamental level, to our knowledge, does not provide interesting predictions about the overall goal of a cell differentiation process. Predictions of this scale would require us to solve the Schrödinger equation for the whole system. Even for simpler systems, at small but almost macroscopic scales, this is very hard and computationally expensive. To make matters worse, this very exact approach might be fundamentally flawed in principle when applied to macroscopic systems (Bolotin 2013). But, also in a very practical sense, large systems are often modelled with approximative methods such as perturbation theory (Fenichel 1979; Gasser and Leutwyler 1984) and WKB[1] approximation (Peterkop 1971) to make the calculations manageable.

Increasing the system size leads us into the realm of molecular dynamics simulations. These require an enormous amount of time, can encompass thousands of particles and quantum mechanical simulations for the active sites of enzymes only (Rommel, Goumans, and Kästner 2011; Warshel 2003). Models with predictive power for macroscopic systems involve approximations and averaging.

Biochemistry is a big part of systems biology by providing insight into the precise sequence of interactions between the molecules within a cell. In some cases the exact biochemical processes are very minute and numerous. In systems biology, they are typically abbreviated into larger, effective interactions. In some cases, prior biochemical knowledge doesn't exist and even very simplified approaches lead to successful causal inference without any knowledge of the mechanism.

However, models that are far removed from our best understanding of the underlying principles of physics and chemistry often comprise large numbers of parameters and consequently exhibit problems such as overfitting. Observations may well be reproduced accurately by such models, but arbitrary behaviour (not found in the system) just as well.

A very coarse approximation or heuristic approaches can also result in very bad fits or the inability to reproduce behaviour once new data is presented. It appears that a useful model needs enough inner structure and a close connection to more fundamental frameworks to have huge predictive power while being as simple as we can reconcile with our observations. Next, we will describe common approaches in systems biology.

A.2. Stochastic and Dynamic Modelling

Models can be fully stochastic, fully deterministic governed by laws of motion, or a mixture like stochastically embedded dynamic models as is the

[1] Wentzel–Kramers–Brillouin

case for deterministic models with noisy measurements.

Although systems in quantum physics are inherently stochastic, these quantum effects are not necessarily required to obtain an accurate description of macropscopic systems. This very much depends on the scale of modelling. It is interesting to note that, at times, biological sensory organs *are* sensitive enough to pick up the stochasticity of elementary particles (Sim et al. 2012). A rather spectacular example of another typical quantum phenomenon is the avian compass, the sensory organ in some birds that enables them to measure magnetic fields. According to Gauger et al. (2011) this biological function exploits entanglement. Perhaps the most relevant intersection of quantum theory and biology lies in the rates of enzymatic reactions. The influence of the *tunnelling* of particles through potential barriers during these reactions is crucial, as it has an impact on the overall *reaction rates* (e.g. Basran, Sutcliffe, and Scrutton 1999; Mills et al. 1997). In *molecular dynamics* simulations the reaction rates are an emergent property, they are a consequence of more fundamental principles. Ordinary differential equation models in systems biology, on the other hand, operate with concentrations of molecules in mind, not particles. They obscure quantum effects and stochasticity from Brownian motion through averaging and *free parametrisation* of *reaction rates*. So, for the most part, these phenomena are neglected, as they are outside of the scope and capability of these models. Since the rates are determined by model fitting, the omission of underlying fundamental physics does not render the modelling fundamentally incorrect.

Statistical approaches can be used to obtain descriptive, stochastic models for the time evolution for the molecule numbers of certain types, e.g. via the chemical master equation (CME) (Gillespie 1976, 1992, 2007). In this case the statistical analysis is applied to pseudo-random events: the chaotic interactions of particles and how they affect molecule numbers over time.

Disregarding individual molecular movement, processes on a cellular level can still appear stochastic (Kramers 1940). Stochastic modelling approaches, such as the Langevin equation (Lemons and Gythiel 1997), have been immensely successful in the modelling of ion channels (Chow and White 1996). The analysis of stochastic models may involve the Fokker-Planck equation (Fokker 1914; Mee and Zweifel 1987) or the CME.

The Gillespie algorithm generates sample trajectories of time continuous reactions that appear as jumps in the number of molecules. The distribution of these trajectories satisfies the CME. Large numbers of molecules can further shift the choice of state variables towards concentrations of substances, described by *discrete time & continuous state space models* (Kramer and Radde

2009; Opgen-Rhein and Strimmer 2007); this class includes autoregressive models (V)AR and autoregressive moving average models ARMA.

The fully discrete Boolean network models (Albert and Othmer 2003; Calzone et al. 2010) depart even more from the underlying biochemistry of intracellular systems but work well with data types such as micro-arrays (Chang 1983; Schena et al. 1995).

Microarrays provide a large quantity of observational readouts but low accuracy for any individual measurement. For these reasons, the readout is best considered as a qualitative *high* or *low* state. This methodology is especially useful for causal inference (Liang, Fuhrman, and Somogyi 1998) and provides information about the possible topology of an interaction network. A special form of causal inference are Bayesian networks (Efron and Tibshirani 1997; Friedman et al. 2000). Yet, once more accurate data needs to be reconciled with the model, it may turn out to be difficult to transform these models into continuous state space and time models.

A very good compromise between the intricacies of interaction mechanisms and practicality with regard to large systems is provided by differential equation models. The averaging of an ensemble of stochastic simulations can be considered a transition between the stochastic and deterministic approach. For large numbers of particles, this type of model is quite justified by the law of large numbers (Slivka and Severo 1970), for very small molecule numbers it can be an oversimplification (El Samad et al. 2005; McAdams and Arkin 1997). In systems biology, the typical case is a large number of reactants and so, both approaches will give similar results (Twycross et al. 2010).

A software package that performs both stochastic or dynamic simulations is COPASI (Hoops et al. 2006). Both techniques are very useful on the macroscopic scale (A. Hastings 1978; Lotka 1920), with the caveat that stochastic simulations are computationally far more expensive. Ordinary differential equations are widespread in many biological fields of research, including cell biology (El-Samad et al. 2005; Trucco 1965a,b), neurobiology (Allken et al. 2014; Hodgkin and Huxley 1952; Jakobsson and Scudiero 1975; Tong, Ghouri, and Taggart 2014), and population dynamics (A. Hastings 1977; Lotka 1920).

ODE models are computationally inexpensive in forward simulations compared to methods involving the CME and Gillespie algorithm but the large numbers of simulations that are needed for MCMC are eventually limiting. The methodology of this approach within systems biology has been applied with great success in a large number of cases such as mitotic cell division (Heinrich et al. 2013), cell proliferation (Hasenauer, Schittler, and

Allgöwer 2012; Schittler, Allgöwer, and De Boer 2013; Schittler, Hasenauer, and Allgöwer 2011), the analysis of circadian rhythms (Leloup, Gonze, and Goldbeter 1999), control of protein secretion (Weber, Hornjik, et al. 2015), and biological switches (Gardner, Cantor, and Collins 2000). This is also true for multicellular systems such as heterogeneous cell populations (Hasenauer, Waldherr, et al. 2011; Spencer et al. 2009).

Since the data has a random measurement error, this approach still requires a statistical framework. This opens our considerations up to Bayesian parameter estimation using Markov chain Monte Carlo simulations.

B. Models

B.1. Test-Model for the mcmc_clib Package

Listing B.1: ODE Model with 26 parameters to test the C software package

```
<?xml version="1.0" ?>
<VectorField Name="ODEmodel11S26P4U"
     Description="A␣model␣for␣testing␣purposes">
<Parameter Name="theta_1" DefaultValue="1.0"/>
<Parameter Name="theta_2" DefaultValue="1.0"/>
<Parameter Name="theta_3" DefaultValue="1.0"/>
<Parameter Name="theta_4" DefaultValue="1.0"/>
<Parameter Name="theta_5" DefaultValue="1.0"/>
<Parameter Name="theta_6" DefaultValue="1.0"/>
<Parameter Name="theta_7" DefaultValue="1.0"/>
<Parameter Name="theta_8" DefaultValue="1.0"/>
<Parameter Name="theta_9" DefaultValue="1.0"/>
<Parameter Name="theta_10" DefaultValue="1.0"/>
<Parameter Name="theta_11" DefaultValue="1.0"/>
<Parameter Name="theta_12" DefaultValue="1.0"/>
<Parameter Name="theta_13" DefaultValue="1.0"/>
<Parameter Name="theta_14" DefaultValue="1.0"/>
<Parameter Name="theta_15" DefaultValue="1.0"/>
<Parameter Name="theta_16" DefaultValue="1.0"/>
<Parameter Name="theta_17" DefaultValue="1.0"/>
<Parameter Name="theta_18" DefaultValue="1.0"/>
<Parameter Name="theta_19" DefaultValue="1.0"/>
<Parameter Name="theta_20" DefaultValue="1.0"/>
<Parameter Name="theta_21" DefaultValue="1.0"/>
<Parameter Name="theta_22" DefaultValue="1.0"/>
<Parameter Name="theta_23" DefaultValue="1.0"/>
<Parameter Name="theta_24" DefaultValue="1.0"/>
<Parameter Name="theta_25" DefaultValue="1.0"/>
<Parameter Name="theta_26" DefaultValue="1.0"/>
<Parameter Name="u1" DefaultValue="0.0"/>
<Parameter Name="u2" DefaultValue="0.0"/>
<Parameter Name="u3" DefaultValue="0.0"/>
<Parameter Name="u4" DefaultValue="0.0"/>
<Expression Name="logistic" Formula="1.0/(1+exp(-t))"/>
```

```
    <Expression Name="U1t" Formula="u1*logistic"/>
    <Expression Name="U2t" Formula="u2*logistic"/>
    <Expression Name="U3t" Formula="u3*logistic"/>
    <Expression Name="U4t" Formula="u4*logistic"/>
39  <Expression Name="S1" Formula="X4+X5"/>
    <Expression Name="S2" Formula="(theta_22)*X9*X7"/>
    <Expression Name="A1" Formula="(theta_1)*X2"/>
    <Expression Name="A2" Formula="(theta_2)*X1"/>
    <Expression Name="A3" Formula="(theta_3)*X3*X2"/>
44  <Expression Name="A4" Formula="(theta_4)*X4"/>
    <Expression Name="A5" Formula="(theta_10)*X4"/>
    <Expression Name="A51" Formula="(theta_10)*X5"/>
    <Expression Name="A6" Formula="(theta_13)*X7"/>
    <Expression Name="A7" Formula="(theta_14)*S1*X6"/>
49  <Expression Name="A8" Formula="(theta_19)*X10*S1"/>
    <Expression Name="A9" Formula="(theta_20)*X8"/>
    <Expression Name="A10" Formula="(theta_22)*X11*X7"/>
    <Expression Name="A11" Formula="(theta_21)*X9"/>
    <StateVariable Name="X1"
54      DefaultInitialCondition="1000.0"
        Formula="A1-A2+(theta_5)*(0.1+S2)-(theta_7)*X1" />
    <StateVariable Name="X2"
        DefaultInitialCondition="1000.0"
        Formula="-A1+A2-(theta_8)*X2" />
59  <StateVariable Name="X3"
        DefaultInitialCondition="1000.0"
        Formula="-A3+A4+(theta_6)+(theta_12)*U2t-(theta_9)*X3" />
    <StateVariable Name="X4"
        DefaultInitialCondition="1000.0"
64      Formula="A3-A4-A5" />
    <StateVariable Name="X5"
        DefaultInitialCondition="0.0"
        Formula="+(theta_11)*U1t-A51" />
    <StateVariable Name="X6"
69      DefaultInitialCondition="1000.0"
        Formula="A6-A7+(theta_15)+(theta_18)*U3t-(theta_16)*X6" />
    <StateVariable Name="X7"
        DefaultInitialCondition="1000.0"
        Formula="-A6+A7-(theta_17)*X7" />
74  <StateVariable Name="X8"
        DefaultInitialCondition="1000.0"
        Formula="+A8-A9-(theta_24)*X8" />
    <StateVariable Name="X9"
        DefaultInitialCondition="1000.0"
79      Formula="+A9-A11" />
    <StateVariable Name="X10"
        DefaultInitialCondition="1000.0"
```

```
        Formula="-A8+A10" />
    <StateVariable Name="X11"
        DefaultInitialCondition="1000.0"
        Formula="-A10+A11+(theta_23)+(theta_26)*U4t-(theta_25)*X11" />
    <Function Name="Y1" Formula="(X4+X5)"/>
    <Function Name="Y2" Formula="(X3+X5)"/>
    <Function Name="Y3" Formula="(X7+X6)"/>
    <Function Name="Y4" Formula="(X6)"/>
    <Function Name="Y5" Formula="(X10+X9+X11+X8)"/>
    <Function Name="Y6" Formula="X8"/>
    </VectorField>
```

B.2. Test-Model for the Comparison of SMMALA Type Algorithms

This is a simplified model of the PKD interactions at the trans Golgi network (Weber, Hornjik, et al. 2015). This following version of the model is closely related to a precursor discussed in Weber, Hasenauer, et al. (2011).

Listing B.2: Medium size model with 6 state variables a scalar input and 9 parameters.

```
<?xml version="1.0" ?>
<VectorField Name="PKD"
    Description="simple␣PKD␣incoherent␣feed␣forward␣model">
<Parameter Name="theta_1" DefaultValue="0" />
<Parameter Name="theta_2" DefaultValue="0" />
<Parameter Name="theta_3" DefaultValue="0" />
<Parameter Name="theta_4" DefaultValue="0" />
<Parameter Name="theta_5" DefaultValue="0" />
<Parameter Name="theta_6" DefaultValue="0" />
<Parameter Name="theta_7" DefaultValue="0" />
<Parameter Name="theta_8" DefaultValue="0" />
<Parameter Name="theta_9" DefaultValue="0" />
<Parameter Name="u" DefaultValue="1" />
<Expression Name="MM" Formula="theta_8␣*␣x_4/(1+x_4)" />
<StateVariable Name="x_1"
    DefaultInitialCondition="0.271366"
    Formula="1␣+␣theta_1*x_6␣-␣x_1" />
<StateVariable Name="x_2"
    DefaultInitialCondition="0.085845"
    Formula="1␣+␣theta_2*x_1␣-␣x_2" />
<StateVariable Name="x_3"
    DefaultInitialCondition="0.346558"
```

```
    Formula="1␣+␣theta_3*x_2␣-␣x_3" />
<StateVariable Name="x_4"
    DefaultInitialCondition="0.424474"
    Formula="1␣+␣theta_4*x_3␣-␣x_4␣-␣u*x_1*MM" />
<StateVariable Name="x_5"
    DefaultInitialCondition="0.322589"
    Formula="1␣+␣theta_5*x_4␣-␣theta_6*x_5␣-␣theta_9*x_5" />
<StateVariable Name="x_6"
    DefaultInitialCondition="0.147937"
    Formula="1␣+␣theta_9*x_5␣-␣theta_9*x_6" />
<Function Name="y_1" Formula="x_1" />
<Function Name="y_2" Formula="x_3" />
<Function Name="y_3" Formula="x_5" />
</VectorField>
```

Listing B.3: Data file for the PKD model in this Section. Note that data contains negative values: although biologically not very plausible, there is nothing that prevents us from using negative values. This occurs for small concentrations and a Gaussian measurement error model.

```
# sampling specific configurations
sample_size=16384
step_size=0.09
acceptance=0.50
# output=./sample/PKD_smmala.double

# experiment specific initial conditions
t0=0.0

# measurement time points
[time]
0.0
5.0
10.0
[/time]

# inputs describe the conditions
# of the system during the experiment
# one line per experiment
[input]
0.4
0.6
0.8
1.0
1.2
2.0
```

```
    [/input]

29  [output]
    1 0 0 0 0 0
    0 0 1 0 0 0
    0 0 0 0 1 0
    [/output]
34
    # rows correspond to entries in «time»
    # columns correspond to different observables (‹Functions› in vfgen)
    # Data Block for Input line 1
    [data]
39  -0.943345        0.762886         0.022923
     1.827489        3.414446         1.754696
     2.420367        5.041516         1.697513
    #
     0.211425        0.452646         0.298842
44   2.435581        4.044996         0.629250
     2.199990        4.330097         2.311937
    #
     0.183554        1.014290        -0.327578
     2.735425        3.870707         1.069117
49   1.403180        3.065056         2.107335
    #
    -0.065283        1.287662         1.100714
     2.687815        4.254431         0.498628
     2.549993        4.815242         0.717132
54  #
    -0.192445        0.105825        -0.074298
     2.059865        2.742548         1.375267
     2.046348        3.632013         0.754942
    #
59  -0.330726        1.178419        -0.364697
     2.885067        3.777156         0.810321
     1.840672        4.741767        -0.120850
    [/data]

64  # standard deviations of data
    # same format as data; not measured Observables
    # have infinite standard deviations
    [sd_data]
    0.600000         0.600000         0.600000
69  0.600000         0.600000         0.600000
    0.600000         0.600000         0.600000
    #
    0.600000         0.600000         0.600000
    0.600000         0.600000         0.600000
```

```
74  0.600000        0.600000        0.600000
    #
    0.600000        0.600000        0.600000
    0.600000        0.600000        0.600000
    0.600000        0.600000        0.600000
79  #
    0.600000        0.600000        0.600000
    0.600000        0.600000        0.600000
    0.600000        0.600000        0.600000
    #
84  0.600000        0.600000        0.600000
    0.600000        0.600000        0.600000
    0.600000        0.600000        0.600000
    #
    0.600000        0.600000        0.600000
89  0.600000        0.600000        0.600000
    0.600000        0.600000        0.600000
    [/sd_data]

    # Gaussian Prior Setup: N(\mu,\Sigma)
94  # \mu
    [prior_mu]
       0.2651763
      -0.1228582
      -0.1897201
99    -1.0223961
       0.0118172
      -0.0075903
       0.6185178
      -0.0878446
104   -0.0900750
    [/prior_mu]

    # \Sigma^{-1}
    [prior_inverse_cov]
109 0.5     0       0       0       0       0       0       0       0
    0       0.5     0       0       0       0       0       0       0
    0       0       0.5     0       0       0       0       0       0
    0       0       0       0.5     0       0       0       0       0
    0       0       0       0       0.5     0       0       0       0
114 0       0       0       0       0       0.5     0       0       0
    0       0       0       0       0       0       0.5     0       0
    0       0       0       0       0       0       0       0.5     0
    0       0       0       0       0       0       0       0       0.5
    [/prior_inverse_cov]
```

B.3. Models Used in the MATLAB Study of RMHMC and SMMALA

The models in this section are shown only for the purposes of this manuscript. The model files we used for the study contained lines for the integration of the state sensitivities and for RMHMC many additional lines for the second order sensitivity. For more detail we refer the reader to the main publication of these results (Kramer, Calderhead, and Radde 2014).

Listing B.4: A very small model with an explicit transformation of the sampling parameters θ_i from logarithmic space.

```
********** MODEL NAME
MAPK
********** MODEL NOTES
exponential parameters
********** MODEL STATES
d/dt(x_01) = - ( (2.0)*exp(theta_01)*(1.0+u_1)^(-1)
                 +exp(theta_02) ) * x_01
             +u_1*exp(theta_01)*(1.0+u_1)^(-1)
             - ( exp(theta_01)*(1.0+u_1)^(-1)
                 -exp(theta_02) ) *x_02
d/dt(x_02) = x_01*exp(theta_01)*(1.0+u_1)^(-1)
             -x_02*exp(theta_02)

x_01(0) = 1.0
x_02(0) = 1.0

********** MODEL PARAMETERS
theta_01 = 0.0
theta_02 = 0.0
u_1 = 0.0
********** MODEL VARIABLES

********** MODEL REACTIONS

********** MODEL FUNCTIONS

********** MODEL EVENTS

********** MODEL MATLAB FUNCTIONS
```

Listing B.5: A mid-sized model with 6 unknown parameters. Model structure taken from the Supplement in Brännmark et al. (2010)

```
********** MODEL NAME
Mma
```

```
    ********** MODEL NOTES
    exponential parameters
    ********** MODEL STATES
 6  d/dt(x_01) = -flux_02-flux_01+flux_03
    d/dt(x_02) = flux_02-flux_04+flux_01
    d/dt(x_03) = flux_05-flux_06

    x_01(0) = 10
11  x_02(0) = 0
    x_03(0) = 0
    ********** MODEL PARAMETERS
    theta_01 = 0.0
    theta_02 = 0.0
16  theta_03 = 0.0
    theta_04 = 0.0
    theta_05 = 0.0
    theta_06 = 0.0
    u_1 = 0.0
21  ********** MODEL VARIABLES
    rho_01 = exp(theta_01)
    rho_02 = exp(theta_02)
    rho_03 = exp(theta_03)
    rho_04 = exp(theta_04)
26  rho_05 = exp(theta_05)
    rho_06 = exp(theta_06)
    ********** MODEL REACTIONS
    flux_01=x_01*rho_01
    flux_02=x_01*u_1*rho_02
31  flux_03=-(-10.0+x_01+x_02)*rho_03
    flux_04=rho_04*x_02
    flux_05=-rho_05*x_02*(-10.0+x_03)
    flux_06=x_03*rho_06
    ********** MODEL FUNCTIONS
36
    ********** MODEL EVENTS

    ********** MODEL MATLAB FUNCTIONS
```

Listing B.6: A fairly large model with 14 unknown parameters. Model structure taken from the Supplement in Brännmark et al. (2010)

```
 1  ********** MODEL NAME
    Mifa
    ********** MODEL NOTES
    exponential parameters
    ********** MODEL STATES
 6  d/dt(x_01) = flux_02+flux_07-flux_01+flux_06
```

```
    d/dt(x_02) = -flux_02+flux_01-flux_03
    d/dt(x_03) = -flux_04-flux_06+flux_03
    d/dt(x_04) = -flux_05+flux_04
    d/dt(x_05) = -flux_09+flux_08
11  d/dt(x_06) = flux_10-flux_11

    x_01(0) = 10
    x_02(0) = 0
    x_03(0) = 0
16  x_04(0) = 0
    x_05(0) = 0
    x_06(0) = 0
    ********** MODEL PARAMETERS
    theta_01 = 0.0
21  theta_02 = 0.0
    theta_03 = 0.0
    theta_04 = 0.0
    theta_05 = 0.0
    theta_06 = 0.0
26  theta_07 = 0.0
    theta_08 = 0.0
    theta_09 = 0.0
    theta_10 = 0.0
    theta_11 = 0.0
31  theta_12 = 0.0
    theta_13 = 0.0
    theta_14 = 0.0
    u_1 = 0.0
    ********** MODEL VARIABLES
36  rho_01 = exp(theta_01)
    rho_02 = exp(theta_02)
    rho_03 = exp(theta_03)
    rho_04 = exp(theta_04)
    rho_05 = exp(theta_05)
41  rho_06 = exp(theta_06)
    rho_07 = exp(theta_07)
    rho_08 = exp(theta_08)
    rho_09 = exp(theta_09)
    rho_10 = exp(theta_10)
46  rho_11 = exp(theta_11)
    rho_12 = exp(theta_12)
    rho_13 = exp(theta_13)
    rho_14 = exp(theta_14)
    ********** MODEL REACTIONS
51  flux_01=x_01*rho_02+u_1*x_01*rho_01
    flux_02=rho_03*x_02
    flux_03=rho_13*x_02
```

```
    flux_04=x_03*rho_14
    flux_05=x_04*(rho_04+x_06*rho_05*(1.0+x_06)^(-1))
56  flux_06=x_03*rho_06
    flux_07=-(-10.0+x_04+x_01+x_03+x_02)*rho_07
    flux_08=-rho_08*(x_03+x_04*rho_09)*(-10.0+x_05)
    flux_09=x_05*rho_10
    flux_10=-rho_11*(-10.0+x_06)*x_05
61  flux_11=x_06*rho_12
    ********** MODEL FUNCTIONS

    ********** MODEL EVENTS

66  ********** MODEL MATLAB FUNCTIONS
```

C. Treatment of Special Cases in Measurement Setups

Here we illustrate our approach when dealing with data that does not conform to simple matrix forms. This can also be inferred from the model files in the previous Chapter.

C.1. Measurement Times Can Be Different between Experiments

If the experiments differ in the measurement time points t_j or the observed molecular species, then the missing values in array y are set to `NA` and the corresponding measurement uncertainty is considered to be infinite to conform to modelling languages which expect full matrices of data. Consider the following (artificial) example:

t	y_1	t	y_2
1 h	1.2(1) M	2 h	13.2(8) M
8 h	1.7(2) M	7 h	16.1(9) M
16 h	2.3(2) M	14 h	17.8(12) M

(C.1)

results in these data matrices:

$$t = (1, 2, 7, 8, 14, 16) \tag{C.2}$$

$$y = \begin{pmatrix} 1.2 & \mathtt{NA} & \mathtt{NA} & 1.7 & \mathtt{NA} & 2.3 \\ \mathtt{NA} & 13.2 & 16.1 & \mathtt{NA} & 17.8 & \mathtt{NA} \end{pmatrix} \tag{C.3}$$

$$\sigma = \begin{pmatrix} 0.1 & \infty & \infty & 0.2 & \infty & 0.2 \\ \infty & 0.8 & 0.9 & \infty & 1.2 & \infty \end{pmatrix}. \tag{C.4}$$

C.2. Log-Normal Measurement Errors and Poor Data

In literature, reported standard deviations can seem exaggerated or otherwise inaccurate. In some cases they are omitted entirely and have to be estimated (crudely) before model fitting can occur.

The results in Kreutz et al. (2007) strongly suggest that the errors in Western blotting are log-normal: $y \sim \log \mathcal{N}(\mu, \sigma)$. The authors suggest to subject the data to a log-transformation, which results in additive Gaussian noise. This is a good plan of action for data with reliable uncertainty quantification. Since the log-normal distribution has positive support, it seems perfectly compatible with measurements of concentrations which are positive by definition and therefore very *plausible*.

However, a major problem arises whenever the data is collected without proper statistical error estimation. When measurements are not repeated often enough, it becomes impossible to obtain a mean *and* its standard deviation.

We have decided to infer the standard deviations σ_{ijk} from the data with some additional assumptions, whenever measurement repetitions $y_{ijk,r}$ ($r = 1, \dots, n_\mathrm{r}$) are available but few $2 < n_\mathrm{r} < 6$.

Data points that are collected by the *same method* are assumed to be subjected to very *similar noise*. If *all* data is collected the same way, we estimate a single scalar uncertainty parameter $\hat{\sigma}$:

$$y_{ijk} = \frac{1}{n_\mathrm{r}} \sum_{r=1}^{n_\mathrm{r}} y_{ijk,r}\,, \tag{C.5}$$

$$\hat{\sigma} = \sqrt{\frac{1}{n_y T n_E} \sum_{ijkr} \frac{(y_{ijk,r} - y_{ijk})^2}{n_\mathrm{r} - 1}}\,. \tag{C.6}$$

This approach assigns the same absolute weight to all data points within an experiment.

One seemingly effective alternative lies in the Bayesian estimation of σ using the model M. We rejected this approach for the sake of simplicity, but more importantly because attempting to fit the model parameters using uncertain observations while also estimating the very uncertainty at the same time seems to be akin to circular reasoning.

In logarithmic space this approach results in very different weighting of the data points. The parameter μ alone determines the median e^μ of the log-normal distribution; in our case $\mu = y_{ijk}$. The variance v however is influenced by both μ and σ:

$$v = (e^{\sigma^2} - 1)e^{2\mu + \sigma^2}\,. \tag{C.7}$$

The consequence of assigning one estimate $\hat{\sigma}$ to all data points is that small measurements are implicitly considered precise while large measured values appear sloppy; the fit will be determined by the smallest measured value whenever the data varies in magnitude.

It seems that for *log-normal error models,* each measurement has to be *repeated often enough* to estimate its individual σ_{ijk} *accurately*.

Bibliography

Albert, Réka and Hans G Othmer (2003). „The topology of the regulatory interactions predicts the expression pattern of the segment polarity genes in Drosophila melanogaster". In: *Journal of theoretical biology* 223.1, pp. 1–18.

Allken, Vaneeda et al. (2014). „The Subcellular Distribution of T-Type Ca2+ Channels in Interneurons of the Lateral Geniculate Nucleus". In: *PloS one* 9.9, e107780.

Alon, Uri (2007). „Network motifs: theory and experimental approaches". In: *Nature Reviews Genetics* 8.6, pp. 450–461.

Anderson, Herbert L. (1986). „Metropolis, Monte Carlo, and the MANIAC". In: *Los Alamos Science* 14, pp. 96–108.

Asmussen, Søren and Peter W. Glynn (2011). „A new proof of convergence of MCMC via the ergodic theorem". In: *Statistics and Probability Letters* 81 (1), pp. 1482–1485.

Basran, Jaswir, Michael J Sutcliffe, and Nigel S Scrutton (1999). „Enzymatic H-Transfer Requires Vibration-Driven Extreme Tunneling". In: *Biochemistry* 38.10, pp. 3218–3222.

Bates, Douglas M. and Donald G. Watts (2007). „QR Decomposition using Householder Transformations". In: *Nonlinear Regression Analysis and Its Applications*. Wiley Series in Probability and Statistics. Wiley. Chap. Appendix 2, pp. 286–289. ISBN: 9780470139004.

Bauer, Christian, Alexander Frink, and Richard Kreckel (2002). „Introduction to the GiNaC framework for symbolic computation within the C++ programming language". In: *Journal of Symbolic Computation* 33.1, pp. 1–12.

Beskos, Alexandros et al. (2013). „Optimal tuning of the hybrid Monte Carlo algorithm". In: *Bernoulli* 19.5A, pp. 1501–1534. DOI: 10.3150/12-BEJ414.

Bolotin, Arkady (2013). *Computational Complexity and the Interpretation of a Quantum State Vector*. Version 1.

Brännmark, Cecilia et al. (2010). „Mass and information feedbacks through receptor endocytosis govern insulin signaling as revealed using a parameter-free modeling framework". In: *J BIOL CHEM* 285.26, pp. 20171–9. DOI: 10.1074/jbc.M110.106849.

Brooks, Steve and Andrew Gelman (1998). „General Methods for Monitoring Convergence of Iterative Simulations". In: *Journal of Computational and Graphical Statistics* 7.4, pp. 434–455. DOI: 10.1080/10618600.1998.10474787.

Brooks, Steve, Andrew Gelman, et al., eds. (2011). *Handbook of Markov Chain Monte Carlo*. Handbooks of Modern Statistical Methods. Chapman & Hall/CRC.

Burnette, W Neal (1981). „Western blotting: electrophoretic transfer of proteins from sodium dodecyl sulfate-polyacrylamide gels to unmodified nitrocellulose and radiographic detection with antibody and radioiodinated protein A". In: *Analytical biochemistry* 112.2, pp. 195–203.

Calderhead, Ben (2011). „Differential Geometric MCMC Methods and Applications". PhD thesis. University of Glasgow.

Calderhead, Ben and Mark Girolami (2009). „Estimating Bayes Factors via Thermodynamic Integration and Population MCMC". In: *Comput. Stat. Data Anal.* 53.12, pp. 4028–4045. ISSN: 0167-9473. DOI: 10.1016/j.csda.2009.07.025.

Calzone, Laurence et al. (2010). „Mathematical Modelling of Cell-Fate Decision in Response to Death Receptor Engagement". In: *PLoS Comput Biol* 6.3, e1000702. DOI: 10.1371/journal.pcbi.1000702.

Chang, Tse-Wen (1983). „Binding of cells to matrixes of distinct antibodies coated on solid surface". In: *Journal of immunological methods* 65.1, pp. 217–223.

Chen, William W., Mario Niepel, and Peter K. Sorger (2010). „Classic and Contemporary Approaches to Modeling Biochemical Reactions". In: *Genes & Development* 24 (17), pp. 1861–1875. DOI: 10.1101/gad.1945410.

Chis, Oana-Teodora, Julio R. Banga, and Eva Balsa-Canto (2011). „Structural Identifiability of Systems Biology Models: A Critical Comparison of Methods". In: *PLoS ONE* 6.11, e27755. DOI: 10.1371/journal.pone.0027755.

Chow, Carson C and John A White (1996). „Spontaneous action potentials due to channel fluctuations". In: *Biophysical Journal* 71.6, pp. 3013–3021.

Costa, Rafael S. et al. (2014). „An extended dynamic model of Lactococcus lactis metabolism for mannitol and 2,3-butanediol production". In: *Mol. BioSyst.* 10 (3), pp. 628–639. DOI: 10.1039/C3MB70265K.

Cowles, Mary Kathryn and Bradley P. Carlin (1996). „Markov Chain Monte Carlo Convergence Diagnostics: A Comparative Review". In: *Journal of the American Statistical Association* 91.434, pp. 883–904. DOI: 10.1080/01621459.1996.10476956.

D'Agostino, Ralph B., Albert Belanger, and Ralph B. D'Agostino Jr. (1990). „A suggestion for using powerful and informative tests of normality". In: *The American Statistician* 44.4, pp. 316–321.

Dutta-Roy, Ranjita et al. (2015). „Ligand-Dependent Opening of the Multiple AMPA Receptor Conductance States: A Concerted Model". In: *PLoS ONE* 10.1, e0116616. DOI: 10.1371/journal.pone.0116616.

Eaton, John W., David Bateman, and Søren Hauberg (2009). *GNU Octave version 3.0.1 manual: a high-level interactive language for numerical computations.* CreateSpace Independent Publishing Platform. ISBN: 1441413006.

Efron, Bradley and Robert Tibshirani (1997). „Improvements on cross-validation: the 632+ bootstrap method". In: *Journal of the American Statistical Association* 92.438, pp. 548–560.

El Samad, Hana et al. (2005). „Stochastic modelling of gene regulatory networks". In: *International Journal of Robust and Nonlinear Control* 15.15, pp. 691–711.

Fenichel, Neil (1979). „Geometric singular perturbation theory for ordinary differential equations". In: *Journal of Differential Equations* 31.1, pp. 53–98.

Fokker, A. D. (1914). „Die mittlere Energie rotierender elektrischer Dipole im Strahlungsfeld". German. In: *Annalen der Physik* 348.5, pp. 810–820. ISSN: 1521-3889. DOI: 10.1002/andp.19143480507.

Francis, J. G. F. (1961). „The QR Transformation A Unitary Analogue to the LR Transformation—Part 1". In: *The Computer Journal* 4.3, pp. 265–271. DOI: 10.1093/comjnl/4.3.265.

Francis, J. G. F. (1962). „The QR Transformation—Part 2". In: *The Computer Journal* 4.4, pp. 332–345.

Friedman, Nir et al. (2000). „Using Bayesian networks to analyze expression data". In: *Journal of computational biology* 7.3-4, pp. 601–620.

Fritsche-Guenther, R. et al. (2011). „Strong negative feedback from Erk to Raf confers robustness to MAPK signalling". In: *MOL SYST BIOL* 7.489.

Gardner, Timothy S, Charles R Cantor, and James J Collins (2000). „Construction of a genetic toggle switch in Escherichia coli". In: *Nature* 403.6767, pp. 339–342.

Gasser, Juerg and Heinrich Leutwyler (1984). „Chiral perturbation theory to one loop". In: *Annals of Physics* 158.1, pp. 142–210.

Gassmann, Max et al. (2009). „Quantifying Western blots: Pitfalls of densitometry". In: *Electrophoresis* 30.11, pp. 1845–1855. ISSN: 1522-2683. DOI: 10.1002/elps.200800720.

Gauger, Erik M et al. (2011). „Sustained quantum coherence and entanglement in the avian compass". In: *Physical Review Letters* 106.4, p. 040503.

Gelman, Andrew, John B. Carlin, et al. (2013). *Bayesian Data Analysis*. Texts in Statistical Science. Chapman & Hall/CRC.

Gelman, Andrew, Gareth O. Roberts, and Walter R. Gilks (1996). „Efficient Metropolis jumping rules". In: *Bayesian Statistics 5*. Ed. by J M Bernardo et al. Vol. 5. Oxford University Press, pp. 599–607.

Gelman, Andrew and D. B. Rubin (1992). „Inference from Iterative Simulation Using Multiple Sequences". In: *Statistical Science* 7.4, pp. 457–472. DOI: 10.1214/ss/1177011136.

Geyer, Charles J. (2011). „Introduction to Markov Chain Monte Carlo". In: *Handbook of Markov Chain Monte Carlo*. Ed. by Steve Brooks et al. Handbooks of Modern Statistical Methods. Chapman & Hall/CRC. Chap. 1, pp. 3–47.

Gillespie, Daniel T (1976). „A general method for numerically simulating the stochastic time evolution of coupled chemical reactions". In: *Journal of computational physics* 22.4, pp. 403–434.

Gillespie, Daniel T (1992). „A rigorous derivation of the chemical master equation". In: *Physica A: Statistical Mechanics and its Applications* 188.1, pp. 404–425.

Gillespie, Daniel T (2007). „Stochastic simulation of chemical kinetics". In: *Annu. Rev. Phys. Chem.* 58, pp. 35–55.

Girolami, Mark and Ben Calderhead (2011). „Riemann manifold Langevin and Hamiltonian Monte Carlo methods". In: *Journal of the Royal Statistical Society: Series B (Statistical Methodology)* 73.2, pp. 123–214. ISSN: 1467-9868. DOI: 10.1111/j.1467-9868.2010.00765.x.

Haario, Heikki, Marko Laine, et al. (2006). „DRAM: Efficient adaptive MCMC". In: *Statistics and Computing* 16, pp. 339–354.

Haario, Heikki, Eero Saksman, and Johanna Tamminen (1998). „An Adaptive Metropolis algorithm". In: *Bernoulli* 7, pp. 223–242.

Hammersley, John Michael and David Christopher Handscomb (1964). *Monte Carlo methods*. Vol. 1. Springer.

Hasenauer, Jan, Daniella Schittler, and Frank Allgöwer (2012). „Analysis and simulation of division- and label-structured population models". In: *Bulletin of Mathematical Biology* 74.11, pp. 2692–2732.

Hasenauer, Jan, Steffen Waldherr, et al. (2011). „Analysis of heterogeneous cell populations: a density-based modeling and identification framework". In: *Journal of Process Control* 21.10, pp. 1417–1425.

Hastings, Alan (1977). „Spatial heterogeneity and the stability of predator-prey systems". In: *Theoretical Population Biology* 12.1, pp. 37–48.

Hastings, Alan (1978). „Global stability in Lotka-Volterra systems with diffusion". In: *Journal of Mathematical Biology* 6.2, pp. 163–168.
Hastings, W. K. (1970). „Monte Carlo Sampling Methods Using Markov Chains and Their Applications". English. In: *Biometrika* 57.1, pp. 97–109. ISSN: 00063444.
Hecker, Michael et al. (2009). „Gene regulatory network inference: data integration in dynamic models—a review". In: *Biosystems* 96.1, pp. 86–103.
Heinrich, Stephanie et al. (2013). „Determinants of robustness in spindle assembly checkpoint signalling". In: *Nature cell biology* 15, pp. 1328–1339. DOI: doi:10.1038/ncb2864.
Hodgkin, Alan L and Andrew F Huxley (1952). „A quantitative description of membrane current and its application to conduction and excitation in nerve". In: *The Journal of physiology* 117.4, p. 500.
Hoops, Stefan et al. (2006). „COPASI–a complex pathway simulator". In: *Bioinformatics* 22.24, pp. 3067–3074.
Huang, C. Y. and J. E. Ferrel (1996). „Ultrasensitivity in the mitogen-activated protein kinase cascade". In: *Proceedings of the National Academy of Sciences of the United States of America* 93, pp. 10078–10083.
Hug, S et al. (2013). „High-dimensional Bayesian parameter estimation: Case study for a model of JAK2/STAT5 signaling". In: *Mathematical Biosciences.*
Jakobsson, E and CARMEN Scudiero (1975). „A transient excited state model for sodium permeability changes in excitable membranes". In: *Biophysical journal* 15.6, pp. 577–590.
Kalman, Rudolf Emil (1970). *Lectures on controllability and observability.* Tech. rep. DTIC Document.
Kazantzis, Nikolaos and Costas Kravaris (1998). „Nonlinear observer design using Lyapunov's auxiliary theorem". In: *Systems & Control Letters* 34.5, pp. 241–247.
Khatri, Purvesh, Marina Sirota, and Atul J. Butte (2012). „Ten Years of Pathway Analysis: Current Approaches and Outstanding Challenges". In: *PLoS Comput Biol* 8.2, e1002375. DOI: 10.1371/journal.pcbi.1002375.
Kramer, Andrei, Ben Calderhead, and Nicole Radde (2014). „Hamiltonian Monte Carlo methods for efficient parameter estimation in steady state dynamical systems". In: *BMC Bioinformatics* 15.1, p. 253. ISSN: 1471-2105. DOI: 10.1186/1471-2105-15-253.
Kramer, Andrei, Jan Hasenauer, et al. (2010). „Computation of the posterior entropy in a Bayesian framework for parameter estimation in biological networks". In: *IEEE International Conference on Control Applications.* (Yoko-

hama, Japan). Part of 2010 IEEE Multi-Conference on Systems and Control, pp. 493–498.

Kramer, Andrei and Nicole Radde (2009). „A Stochastic Framework for Noise Separation in Dynamic Models of Intracellular Networks". In: *Proc. CASYS'09*. Ed. by Daniel M. Dubois. 9. AIP, pp. 68–73.

Kramer, Andrei and Nicole Radde (2010). „Towards experimental design using a Bayesian framework for parameter identification in dynamic intracellular network models". In: *Procedia Computer Science*. ICCS 2010. (Amsterdam). Ed. by J. Liebowitz. Vol. 1. 1, pp. 1645–1653. DOI: 10.1016/j.procs.2010.04.184.

Kramer, Andrei, Vassilios Stathopoulos, et al. (2014). „mcmc_clib – an advanced MCMC sampling package for ode models". In: *Bioinformatics* 30 (20), pp. 2991–2992. DOI: 10.1093/bioinformatics/btu429.

Kramers, Hendrik Anthony (1940). „Brownian motion in a field of force and the diffusion model of chemical reactions". In: *Physica* 7.4, pp. 284–304.

Kreutz, C. et al. (2007). „An error model for protein quantification". In: *Bioinformatics* 23.20, pp. 2747–2753. DOI: 10.1093/bioinformatics/btm397.

Kublanovskaya, Vera N. (1963). „On some algorithms for the solution of the complete eigenvalue problem," in: *Computational Mathematics and Mathematical Physics* 1.3, pp. 637–657.

Leimkuhler, Benedict and Sebastian Reich (2004). *Simulating hamiltonian dynamics*. Vol. 14. Cambridge University Press, p. 85. 381 pp.

Leloup, Jean-Christophe, Didier Gonze, and Albert Goldbeter (1999). „Limit cycle models for circadian rhythms based on transcriptional regulation in Drosophila and Neurospora". In: *Journal of Biological Rhythms* 14.6, pp. 433–448.

Lemons, Don S. and Anthony Gythiel (1997). „Paul Langevin's 1908 paper "On the Throry of Brownian Motion", «Sur la théorie du mouvement brownien», C. R. Acad. Sci. (Paris) 146, 530–533 (1908)". In: *American Journal of Physics* 65.11, pp. 1079–1081. DOI: http://dx.doi.org/10.1119/1.18725.

Li, Chen et al. (2010). „BioModels Database: An enhanced, curated and annotated resource for published quantitative kinetic models." In: *BMC Systems Biology* 4, p. 92.

Liang, Shoudan, Stefanie Fuhrman, Roland Somogyi, et al. (1998). „Reveal, a general reverse engineering algorithm for inference of genetic network architectures". In: *Pacific symposium on biocomputing*. Vol. 3. 3, pp. 18–29.

Lotka, Alfred J (1920). „Analytical note on certain rhythmic relations in organic systems“. In: *Proceedings of the National Academy of Sciences of the United States of America* 6.7, p. 410.

Madras, Neal and Dana Randall (2002). „Markov Chain Decomposition for Convergence Rate Analysis“. In: *The Annals of Applied Probability* 12.2, pp. 581–606.

Markov, Andrey Andreyevich (1906). „Extension of the law of large numbers to dependent events“. In: *Bull. Soc. Phys. Math. Kazan* 2.15, pp. 155–156.

McAdams, Harley and Adam Arkin (1997). „Stochastic mechanisms in gene expression“. In: *Proceedings of the National Academy of Sciences* 94.3, pp. 814–819.

Mee, CVM van der and PF Zweifel (1987). „A Fokker-Planck equation for growing cell populations“. In: *Journal of mathematical biology* 25.1, pp. 61–72.

Metropolis, Nicholas et al. (1953). „Equation of State Calculations by Fast Computing Machines“. In: *J CHEM PHYS* 21.6, pp. 1087–1092. DOI: 10.1063/1.1699114.

Mills, G et al. (1997). „Generalized path integral based quantum transition state theory“. In: *Chemical physics letters* 278.1, pp. 91–96.

Mira, Antonietta (2001). „On Metropolis-Hastings algorithms with delayed rejection“. In: *Metron - International Journal of Statistics* 3-4, pp. 231–241.

Möller, Yvonne et al. (2014). „EGFR-targeted TRAIL and a Smac mimetic synergize to overcome apoptosis resistance in KRAS mutant colorectal cancer cells“. In: *PloS one* 9.9, e107165. ISSN: 1932-6203. DOI: 10.1371/journal.pone.0107165.

Opgen-Rhein, Rainer and Korbinian Strimmer (2007). „Learning causal networks from systems biology time course data: an effective model selection procedure for the vector autoregressive process“. In: *BMC Bioinformatics* 8.Suppl 2, S3. ISSN: 1471-2105. DOI: 10.1186/1471-2105-8-S2-S3.

Parzen, Emanuel (1962). „On Estimation of a Probability Density Function and Mode“. In: *Ann. Math. Statist.* 33.3, pp. 1065–1076. DOI: 10.1214/aoms/1177704472.

Peterkop, R (1971). „WKB approximation and threshold law for electron-atom ionization“. In: *Journal of Physics B: Atomic and Molecular Physics* 4.4, p. 513.

Plummer, Martyn et al. (2006). „CODA: convergence diagnosis and output analysis for MCMC“. In: *R News* 6.1, pp. 7–11.

Pokhilko, Alexandra, Paloma Mas, and Andrew Millar (2013). „Modelling the widespread effects of TOC1 signalling on the plant circadian clock and its outputs". In: *BMC Systems Biology* 7.1, p. 23. ISSN: 1752-0509. DOI: 10.1186/1752-0509-7-23.

R Development Core Team (2008). *R: A Language and Environment for Statistical Computing*. R Foundation for Statistical Computing. Vienna, Austria. ISBN: 3900051070.

Radde, Nicole (2010). „Fixed point characterization of biological networks with complex graph topology". In: *Bioinformatics* 26.22, pp. 2874–2880. DOI: 10.1093/bioinformatics/btq517.

Radde, Nicole and Jonas Offtermatt (2014). „Convergence of posteriors for structurally non-identified problems using results from the theory of inverse problems". In: *Journal of Inverse and Ill-posed Problems* 22.2, pp. 251–276.

Radford, M. Neil (2011). „Handbook of Markov Chain Monte Carlo". In: ed. by Steve Brooks et al. Handbooks of Modern Statistical Methods. Chapman & Hall/CRC. Chap. 5, pp. 113–162.

Rami, M. Ait, C. H. Cheng, and C. de Prada (2008). „Tight Robust interval observers: an LP approach". In: *Proceedings of the 47th IEEE Conference on Decision and Control*. (Cancun, Mexico).

Rao, C. Radhakrishna (1945). „Information and accuracy attainable in the estimation of statistical parameters." In: *Bull. Calc. Math. Soc.* 37, pp. 81–91.

Rao, C. Radhakrishna (1992). „Information and accuracy attainable in the estimation of statistical parameters." In: *Breakthroughs in Statistics*. Ed. by S. Kotz and N. L. Johnson. Vol. 1. Perspectives in Statistics. Springer. Chap. 8, pp. 243–245.

Renart, Jaime and Ignacio V Sandoval (1984). „Western blots". In: *Methods in enzymology* 104, pp. 455–460.

Roberts, Gareth O. and Jeffrey S. Rosenthal (2001). „Optimal scaling for various Metropolis-Hastings algorithms". In: *Statistical Science* 16.4, pp. 351–367. DOI: 10.1214/ss/1015346320.

Roberts, Gareth O. and Jeffrey S. Rosenthal (2004). „General state space Markov chains and MCMC algorithms". In: *Probability Surveys* 1, pp. 20–71. DOI: http://dx.doi.org/10.1214/154957804100000024.

Roberts, Gareth O. and Richard L. Tweedie (1996). „Exponential Convergence of Langevin Distributions and Their Discrete Approximations". In: *Bernoulli* 2.4, pp. 341–363.

Rommel, Judith B., T. P. M. Goumans, and Johannes Kästner (2011). „Locating Instantons in Many Degrees of Freedom". In: *Journal of Chemical Theory and Computation* 7.3, pp. 690–698. DOI: 10.1021/ct100658y.

Rosenblatt, Murray (1956). „Remarks on Some Nonparametric Estimates of a Density Function". In: *Ann. Math. Statist.* 27.3, pp. 832–837. DOI: 10.1214/aoms/1177728190.

El-Samad, HJCM et al. (2005). „Surviving heat shock: control strategies for robustness and performance". In: *Proceedings of the National Academy of Sciences of the United States of America* 102.8, pp. 2736–2741.

Schena, Mark et al. (1995). „Quantitative monitoring of gene expression patterns with a complementary DNA microarray". In: *Science* 270.5235, pp. 467–470.

Schittler, Daniella, Frank Allgöwer, and Rob De Boer (2013). „A new model to simulate and analyze proliferating cell populations in BrdU labeling experiments". In: *BMC Systems Biology* 7.Suppl 1, S4. ISSN: 1752-0509. DOI: 10.1186/1752-0509-7-S1-S4.

Schittler, Daniella, Jan Hasenauer, and Frank Allgöwer (2011). „A generalized population model for cell proliferation: Integrating division numbers and label dynamics". In: *TICSP series* 57, pp. 165–168.

Schmidt, Henning and Mats Jirstrand (2006). „Systems Biology Toolbox for MATLAB: a computational platform for research in systems biology". In: *Bioinformatics* 22.4, pp. 514–515. DOI: 10.1093/bioinformatics/bti799.

Sim, Nigel et al. (2012). „Measurement of Photon Statistics with Live Photoreceptor Cells". In: *Phys. Rev. Lett.* 109 (11), p. 113601. DOI: 10.1103/PhysRevLett.109.113601.

Skovgaard, Lene Theil (1981). technical report 167. Stanford, California: Standford University, Department of Statistics.

Slivka, John and Norman C Severo (1970). „On the strong law of large numbers". In: *Proceedings of the American Mathematical Society*, pp. 729–734.

Spencer, Sabrina L et al. (2009). „Non-genetic origins of cell-to-cell variability in TRAIL-induced apoptosis". In: *Nature* 459.7245, pp. 428–432.

Taylor, Sean C., Thomas Berkelman, et al. (2013). „A Defined Methodology for Reliable Quantification of Western Blot Data". English. In: *Molecular Biotechnology* 55.3, pp. 217–226. ISSN: 1073-6085. DOI: 10.1007/s12033-013-9672-6.

Taylor, Sean C. and Anton Posch (2014). „The design of a quantitative western blot experiment". In: *BioMed research international* 2014. DOI: 10.1155/2014/361590.

Tong, Wing-Chiu, Iffath Ghouri, and Michael J Taggart (2014). „Computational modeling of inhibition of voltage-gated Ca channels: identification of different effects on uterine and cardiac action potentials". In: *Clinical and Translational Physiology* 5, p. 399.

Trucco, E. (1965a). „Mathematical models for cellular systems the von Foerster equation. Part I". English. In: *The bulletin of mathematical biophysics* 27.3, pp. 285–304. ISSN: 0007-4985. DOI: 10.1007/BF02478406.

Trucco, E. (1965b). „Mathematical models for cellular systems. The von foerster equation. Part II". English. In: *The bulletin of mathematical biophysics* 27.4, pp. 449–471. ISSN: 0007-4985. DOI: 10.1007/BF02476849.

Twycross, Jamie et al. (2010). „Stochastic and deterministic multiscale models for systems biology: an auxin-transport case study". In: *BMC Systems Biology* 4.1, p. 34. ISSN: 1752-0509. DOI: 10.1186/1752-0509-4-34.

Vehlow, Corinna et al. (2012). „Uncertainty-aware visual analysis of biochemical reaction networks". In: *IEEE Symposium on Biological Data Visualization(Biovis)*, pp. 91–98.

Vogel, Christine and Edward M Marcotte (2012). „Insights into the regulation of protein abundance from proteomic and transcriptomic analyses". In: *Nature Reviews Genetics* 13.4, pp. 227–232.

Warshel, Arieh (2003). „Computer Simulations of Enzyme Catalysis: Methods, Progress, and Insights". In: *Annual Review of Biophysics and Biomolecular Structure* 32.1. PMID: 12574064, pp. 425–443. DOI: 10.1146/annurev.biophys.32.110601.141807.

Weber, Patrick, Jan Hasenauer, et al. (2011). „Parameter estimation and identifiability of biological networks using relative data". In: *Proceedings of the 18th IFAC World Congress.* (Università Cattolica del Sacro Cuore, Milano, Italy). 18. IFAC, pp. 11648–11653.

Weber, Patrick, Mariana Hornjik, et al. (2015). „A computational model of PKD and CERT interactions at the trans-Golgi network of mammalian cells." eng. In: *BMC Syst Biol* 9.1, p. 147. DOI: 10.1186/s12918-015-0147-1.

Weber, Patrick, Andrei Kramer, et al. (2012). „Trajectory-oriented Bayesian experiment design versus Fisher A-optimal design: an in depth comparison study". In: *Bioinformatics* 28.18, pp. i535–ii541. DOI: 10.1093/bioinformatics/bts377.

Weckesser, Warren (2008). „VFGEN: A Code Generation Tool". In: *JNAIAM* 3.1-2, pp. 151–165.

Williams, Thomas et al. (1986–2014). *gnuplot, a portable command-line driven graphing utility.* Version 4.6.6. URL: gnuplot.info (visited on 2014).

Wolff, Ulli (2004). „Monte Carlo errors with less errors". In: *COMPUT PHYS COMMUN* 156.2, pp. 143–153. ISSN: 0010-4655. DOI: 10.1016/S0010-4655(03)00467-3.